Respiration and Phosphorylation
of Bacteria

Respiration and Phosphorylation of Bacteria

Nina S. Gel'man, Marina A. Lukoyanova, and Dmitrii N. Ostrovskii

Laboratory of Evolutionary Biochemistry
A. N. Bakh Institute of Biochemistry
Academy of Sciences of the USSR, Moscow

Translated from Russian

Translation Editor
Gifford B. Pinchot
Department of Biology and McCollum-Pratt Institute
The Johns Hopkins University
Baltimore, Maryland

℗ SPRINGER SCIENCE+BUSINESS MEDIA, LLC
1967

Nina Samoilovna Gel'man, Doctor of Biological Sciences, earned her degree at the Department of Plant Biochemistry of the M. V. Lomonosov Moscow State University. She has been a research worker at the A. N. Bakh Institute of Chemistry since 1946, where she is presently on the staff of the Laboratory of Evolutionary Biochemistry, headed by A. I. Oparin, involved in the problems of molecular organization of enzyme systems, as are Marina A. Lukoyanova and Dmitrii N. Ostrovskii, both Candidates of Biological Sciences.

The original Russian text, published by Nauka Press, Moscow, in 1966, has been revised and updated for the American edition.

Нина Самойловна Гельман,
Марина Артуровна Лукоянова,
Дмитрий Николаевич Островский.

Дыхательный аппарат бактерий

DYKHATEL'NYI APPARAT BAKTERII

Library of Congress Catalog Card Number 66-26220

© *1967 Springer Science+Business Media New York*
Originally published by Plenum Press 1967
Softcover reprint of the hardcover 1st edition 1967

ISBN 978-1-4899-5528-9   ISBN 978-1-4899-5526-5 (eBook)
DOI 10.1007/978-1-4899-5526-5

Foreword

This book is obviously of great interest to bacteriologists and all those interested in bacterial metabolism and especially to those interested in bacterial energy metabolism. It will also appeal to a much wider audience for two reasons. First, bacteria because of their primitive position on the evolutionary scale can't help but tell us a great deal about how life began on earth. Second, the use of bacteria in biochemical studies has often, in the past, opened up new areas and provided new knowledge that was later confirmed in mammalian tissues. There is every indication that this will continue to be true in this field.

This book is particularly timely and important in that it provides the only thorough and up-to-date attempt to collect and evaluate the literature on bacterial respiration and oxidative phosphorylation. While there have been a great many reviews of mitochondrial metabolism this is the only one that covers the bacterial literature with any degree of completeness. Making these data available in one place is a very valuable service, which not only provides insight into bacterial mechanisms but also points up similarities and differences between bacterial and mitochondrial structure and metabolism. This book will certainly provide a stimulus for new and exciting research.

Gifford B. Pinchot

Baltimore
June, 1967

Preface to the American Edition

This book is an attempt to sum up the literature on respiration and phosphorylation in bacteria. The topic of discussion is the morphology and molecular organization of the membranous structures, the enzymatic composition of the respiratory chain, the mechanism of oxidative phosphorylation, and possible pathways of evolution of the respiratory apparatus.

Since the progress of research in this field is very rapid there is a danger that the state of the problem as given in the book may lag behind the information which has appeared by the time that the book is published. Nevertheless, we hope that the book will be of use to research workers in the field of respiration, bacterial energetics, and the molecular organization of bacterial membranes.

We take this opportunity of thanking the authors whose diagrams and tables are given in the book.

The Authors

Preface to the Seventh Edition

Contents

Introduction

Intensive research over the last ten years has shown that many of the metabolic pathways, enzyme systems, and intermediate products common to all modern organisms are also found in bacteria, thus confirming the concept of the common origin of all forms of life in earth (Oparin, 1924, 1957, 1960). Although the common nature of several metabolic processes has weakened the sharp distinction between bacteria and highly organized cells, bacteria still occupy a special position among living creatures as regards morphology and biochemical characteristics. Bacteria are unique and extremely interesting organisms. They cannot be regarded as ordinary cells—like animal or plant cells, only smaller. The special nature of bacteria is so pronounced that they sometimes appear to be free-living nuclei or mitochondria.

An interesting feature which distinguishes bacteria from multi-cellular organisms is their ability to develop in the partial or complete absence of air. This ability indicates the antiquity of their origin and, perhaps, even their association with the anaerobic period of the earth's atmosphere (Gaffron, 1964).

The comparatively simple cellular organization and metabolism of bacteria and their ability to live anaerobically suggest that they lie somewhere at the base of the evolutionary ladder. Hence, investigations of bacteria can contribute to theories of the evolution of cell structure and function on the basis of a comparison of bacteria with highly organized cells.

Oparin (1957, 1960, 1966) has suggested that the remote ancestors of present-day organisms can be pictured as blobs of protoplasm surrounded by a single lipoprotein membrane. This membrane had to perform several functions, which were later distributed among the different cell organelles. In particular, this single outer membrane of the earliest aerobic cells must have carried the enzymes of the respira-

1

tory chain and oxidative phosphorylation. The situation is rather similar in the most simply organized organisms living today, such as, for instance, the smallest living creatures (down to 0.1 μ in diameter), belonging to the group of pleuropneumonia-like or coccoid bacteria (Pirie, 1964; Gale, 1962). Hence, an investigation of the structure and physiology of bacteria, particularly small forms, is of very great interest for clarification of the central problems of evolutionary biochemistry. For instance, the study of bacterial nucleic acids can throw light on the origin and development of the genetic apparatus, since the genetic apparatus of bacteria is similar in its abilities to that of the smallest living organisms. The lipoprotein membranes of bacteria provide an abundance of material for the study of poly-functionality—the combined performance of several processes such as photosynthesis, respiration, active ion transfer, etc., by one struc-ture.

In accordance with the widely accepted concept of "biochemi-cal" unity, one of the most significant features of the living state is the continuous and directed movement of electrons in the cell (Kluyver and Van Niel, 1959). The formation of phosphorylated high-energy compounds is the main purpose of this process, which is common to all living things.

Electron transport and oxidative phosphorylation are effected in the respiratory apparatus, which is an assembly of enzyme systems, spatially organized in the structural elements of the cell. In the cells of animals, plants, and most microorganisms, this apparatus consists of a system of enzymes localized in the mitochondria. The cells of bacteria and blue-green algae have enzyme systems similar in compo-sition to those of all other cells, but they possess no mitochondria; hence, the structure of the respiratory apparatus and the spatial organization of the enzymes of electron transport and oxidative phos-phorylation are different. The system of enzymes responsible for the transfer of electrons to oxygen is called "the respiratory chain" or "the electron-transport chain." We prefer the first name, since it expresses the nature of the process more concisely.

Our concepts of the nature and functions of the respiratory chain have been obtained mainly from mitochondria of animal tissues. According to Chance and Williams (1956), the respiratory chain of mitochondria has three main functions: (1) to transport electrons from the substrate to oxygen and, in particular, to maintain the necessary level of oxidized nicotinamide adenine dinucleotide within the cell; (2) to provide at least three sites for energy conservation, where ADP is converted to ATP; and (3) to regulate metabolism in accordance with the levels of control substances (e.g., ADP, NAD, or hormones).

The respiratory chain of mitochondria includes NADH dehydrogenase, succinic dehydrogenase, and cytochromes of groups *A*, *B*, and *C*. These enzymes are assembled in a definite order in the structure of the mitochondrial lipoprotein membranes. Ubiquinone appears to be a component of the system.

In contrast to the situation for animal and plant cells, the respiratory chain of bacteria has not been adequately investigated, although there is now a large body of facts which show that, in the aerobic and facultative anaerobic bacteria that have been investigated, the composition of the respiratory chain is similar in many respects to that of other cells. Another similarity is that the respiratory chain of bacteria is associated with membranous elements of the cell, i.e., is spatially organized. Yet there are no mitochondria, and oxidative phosphorylation in preparations of bacterial membranes is very inefficient. The regulation of oxidative metabolism in bacteria is very inferior to that in mitochondria. There is no respiratory control mechanism in bacteria.[1]

The aim of this book is to characterize the respiratory apparatus of bacteria, i.e., the composition of the enzyme chains and their structural organization. We were interested in the special features of the enzyme composition (the assortment and properties of the components, their number, behavior in regard to inhibitors, methods of fractionation and purification) and the molecular organization of the respiratory chain (the ultrastructure of bacterial membranes, the composition of lipoprotein complexes of membranes, and the role of lipids in maintaining the structure of the respiratory chain and in enzyme activity). In the treatment of this problem we have tried to bring out the similarity and differences between the respiratory apparatus of bacteria and mitochondria, structures with similar functions, which are much more advanced morphologically and evolutionally younger.

[1] Editor's note: This statement is no longer true. See: Ishikawa, S., and Lehninger, A. L., *J. Biol. Chem.* 237, 2401, 1962; and Scocca and Pinchot, *Fed. Proc.* 24, 544, 1965.

Chapter I

Membranous Structures of Bacteria

STRUCTURAL FEATURES OF THE BACTERIAL CELL

Most bacteria are extremely minute organisms. Their average size is about 5 μ (Ierusalimskii, 1963; Zarnea, 1963; Morowitz and Tourtellotte, 1964), and the smallest, the coccoid bacteria for instance, have a diameter of only 0.5–1 μ. Thus, cocci are a thousand times smaller than the average animal cell and are smaller even than such cell organelles as mitochondria, which can attain a length of 10 μ with a diameter of 0.5 μ. (Marr, 1960). A very significant index for illustration of the size of bacteria is the number of free hydrogen ions (H^+) at pH 7 in a volume of liquid equal to the volume of the cell. In this case the cell of a coccus contains 5–50 hydrogen ions (Chen and Cleverdon, 1962). This does not mean that there are always 5–50 H^+ ions in a cell in which the pH of the cytoplasm is 7. Different metabolic processes can alter the H^+ concentration very considerably for certain short intervals of time, but the over-all effect due to these variations corresponds to the H^+ concentration at pH 7. The small size of the bacterial cell is of very great significance for the biochemical processes occurring in it. The size of the cell is particularly important when it is reduced to a few tenths of a micron, only slightly exceeding the theoretically permissible minimum (0.05 μ), as in the case of some forms of mycoplasma (diameter 0.1 μ) (Pirie, 1964; Morowitz *et al.,* 1962; Morowitz and Tourtellotte, 1964).

Calculations show that the bacterial cell can contain a very limited number of protein and nucleic acid molecules. This is reflected in the specificity of bacterial metabolism. For instance, Bresler (1963) points out that the DNA of *Escherichia coli* contains about 1 million units of information. This is sufficient for the synthesis of only 2000 different protein molecules, while *Dialister pneumosintes* can synthesize only 600 types of protein molecules (Chen and Cleverdon, 1962), whereas the DNA of the animal cell carries 1000 times more information. It has been suggested (Gale, 1959) that bacterial proteins must each perform several functions, since he estimates that there are only 400,000 to 500,000 protein molecules per *Staphylococcus aureus* cell.

It has been shown that polyfunctionality is a feature of the few membranous structures (the cytoplasmic membrane, for instance), which must each carry several enzyme systems. These systems are localized in specialized organelles in higher organisms.

The wall of the bacterial cell is about 20% of its weight (27–38% in *Streptococcus faecalis*), but there is no evidence yet that it takes any active part in cell metabolism (Salton, 1964). The relatively large dimensions of the cell wall are attributed simply to the need for mechanical protection of the protoplast (Weidel and Pelzer, 1964). The lipoprotein cytoplasmic membrane plays a much greater role in the life of the bacterial cell. In some bacteria it is the only membranous structure of the cell. It is of interest that the ratio of the area of the cytoplasmic membrane to the volume of the cell is ten times the corresponding value for the animal cell.

The nuclear apparatus of bacteria differs greatly from that of higher organisms. Bacteria do not have a typical nucleus with a distinct membrane and nucleolus (Brieger, 1963; Stanier, 1964). The DNA of bacteria forms a "nuclear body" or "zone" (Van Tubergen and Setlow, 1961; Juhacz, 1961), which lacks a lipoprotein membrane, but can be isolated in fairly pure form (Godson and Butler, 1962; Spiegelman *et al.,* 1958; Echlin and De Lamater, 1962). Specialists believe that this type of nuclear apparatus—a compact endosome according to Dillon's terminology (1962)—is intermediate between the apparatus of blue-green algae, on one hand, and that of protozoa, on the other, although some investigators (Stanier, 1964) put bacteria into an independent evolutionary series.

The photosynthetic apparatus in photosynthetic bacteria has a distinctive structure. The bacteriochlorophyll in them is usually contained in spherical granules, or chromatophores (diameter about 600 Å), which are small in comparison with chloroplasts (Kamen, 1963; Marr, 1960). There are other forms of organization of the photosynthetic apparatus. As distinct from plant chloroplasts, the chromatophores do not contain the enzymes required for CO_2 fixation. Morphologically, as Boatman (1964) recently showed for *Rhodospirillum rubrum* and Cohen-Bazire *et al.* (1964) for *Chlorobium sp.,* the chromatophores are formed by invagination of the cytoplasmic membrane.

Until very recently, no internal membranes or membranous structures resembling the endoplasmic reticulum or mitochondria had been found in the cytoplasm of bacteria (Bradfield, 1956; Luria, 1960). It was believed that the ribosomes lay freely in the cytoplasm and that the respiratory enzymes were localized in the cytoplasmic membrane—the only membrane which could be observed in the bacterial cell (Hughes, 1962). The first reports of membranous structures in bacteria were met with disbelief and were even disputed. However, the presence of membranous structures of various kinds in most bacteria can now be regarded as proven, and attention is being concentrated on their function. If they play the part of an endoplasmic reticulum, they will carry the ribosomes and resemble the canals and cisternae of the reticulum, but will not contain the phosphorylating respiratory chain. If they are equivalent to mitochondria, their structure should show some elements of similarity with that of mitochondria, and they will have to carry the respiratory chain, i.e., the cytochromes and dehydrogenases. It is obvious that obligate anaerobes, which lack a respiratory chain, will not contain membranous structures.

The membranous structures in bacteria are of interest in the study of their respiratory apparatus (structure, molecular organization, and function). For comparison we will later mention information relating to the structure and function of mitochondria, particularly the mitochondria of the mammalian heart, which have been the most thoroughly investigated.

Great progress in the technique of obtaining ultrathin sections and the development of special fixing and staining methods led to the discovery of membranous structures inside the bacterial cell. Representatives of various groups of bacteria and of related microorganisms, such as actinomycetes and blue-green algae, have been investigated. In most investigations, intact cells were the object of study, but the membranous structures have been isolated from some bacteria, and their structure and composition investigated.

The membranous structures investigated in intact cells will be described separately for each bacterial species studied. In view of the great polymorphism of the membranous structures and the impossibility of reproducing numerous photomicrographs we have attempted

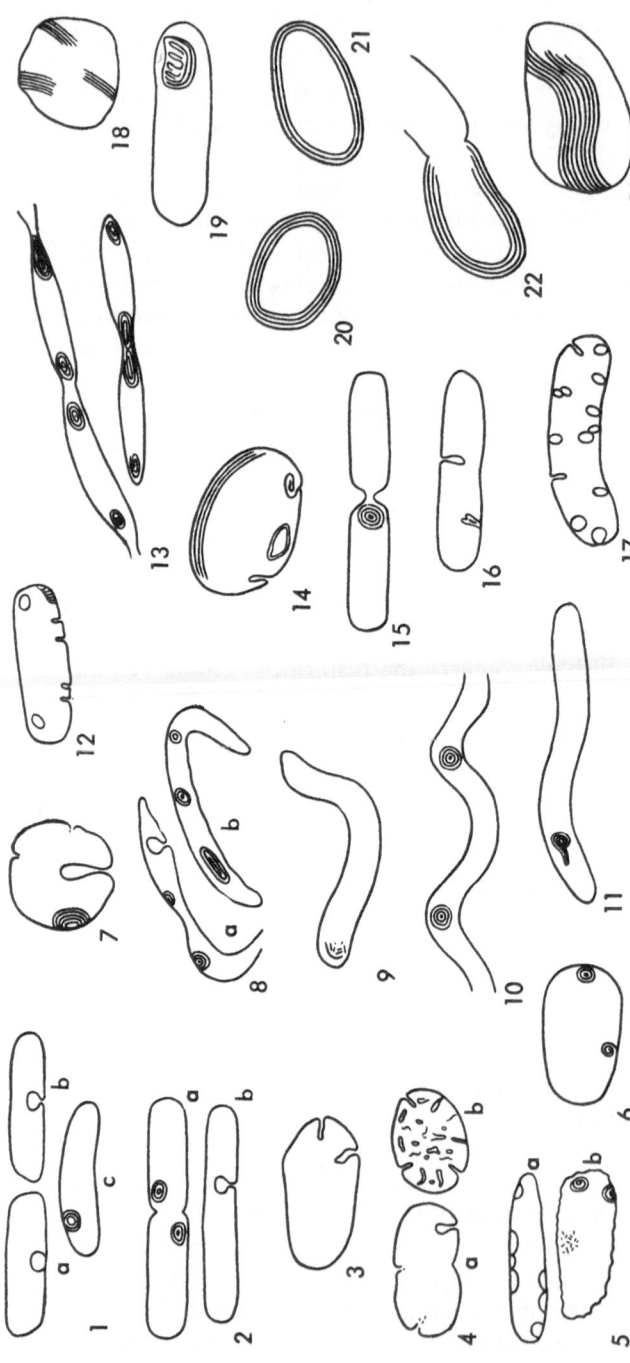

Fig. 1. Membranous structures in gram-negative bacteria. 1. *Escherichia coli* (*a*—Niklowitz, 1958; Biryuzdva, 1960; *b*—Kran, 1962; *c*—Vanderwinkel and Murray, 1962); 2. *Spirillum serpens* (*a*—Vanderwinkel and Murray, 1962; *b*—Murray and Birch-Anderson, 1963); 3. *Acetobacter suboxydans* (Claus and Roth, 1964); 4. *Azotobacter sp.* (*a*—Wyss *et al.*, 1961; *b*—Pangborn *et al.*, 1962; Van Iterson, 1963); 5. *Proteus. vulgaris* (*a*—Nermut and Rýc, 1964; *b*—Van Iterson and Leene, 1964b); 6. *Rhizobium sp.* (Dart and Mercer, 1963b; Dixon, 1964); 7. *Neisseria gonorrheae* (Fitz-James, 1964a); 8. *Treponema Reiter* (*a*—Ryter and Pillot, 1963; *b*—Kawata and Inoue, 1964); 9. *Treponema microdentium* (Listgarten *et al.*, 1963); 10. *Leptospira pomona* (Ritchie and Ellinghausen, 1965); 11. *Borrelia recurrentis* (Ludvik, 1964); 12. *Brucella abortus* S and R (De Petris *et al.*, 1964); 13. *Caulobacter bacteroides* (Stove Poindexter and Cohen-Bazire, 1964); 14. *Hyphomicrobium sp.* (Conti and Hirsch, 1965); 15. *Bacteroides ruminicola* (Bladen and Waters, 1963); 16. *Fusobacterium polymorphum* (Takagi and Uejama, 1963); 17. *Rhodospirillum rubrum* (Cohen-Bazire and Kunizawa, 1963; Boatman, 1964); 18. *Rhodospirillum molishianum* (Giesbrecht and Drews, 1962); 19. *Chlorobium sp.* (Cohen-Bazire *et al.*, and Watson, 1965); 20. *Rhodomicrobium vannielii* (Boatman and Douglas, 1961; Conti and Hirsch, 1965); 21. *Nitrosomonas europea* (Murray, 1963; Murray and Watson, 1965); 22. *Nitrobacter agilis* (Murray, 1963; Murray and Watson, 1965); 23. *Nitrosocystis oceanus* (Murray, 1963; Murray and Watson, 1965).

to schematize their structure (see Fig. 1 for gram-negative and Fig. 2 for gram-positive bacteria). Classification by Gram staining is a rather formal approach, particularly since the nature of the reaction has not yet been explained (Salton, 1964). It is known, however, that division of bacteria into these two groups is correlated in some way with some physiological and biochemical characters and with structural features of the cell wall. All spore-forming bacteria are gram-positive, whereas gram-negative bacteria never produce spores. Autotrophic bacteria are gram-negative. For some unknown reason, the cells of gram-negative species usually contain ubiquinones, while gram-positive cells contain naphthoquinones. An examination of the membranous structures also shows a definite correlation with Gram staining.

To explain the diagrams shown in Figs. 1 and 2, we will recall that absolutely all bacterial cells have a similarly constructed cytoplasmic membrane, 75–80 Å thick, enveloping the protoplast. Structurally, internal membranous formations differ considerably in different bacteria and in some bacteria they are not found at all. The structure of the cytoplasmic membrane was first shown by means of a special fixing technique by Kellenberger and Ryter (1958), using *E. coli*. The ultrastructure and molecular organization of the cytoplasmic membrane and membranous structures will be discussed in special sections. For convenience in Figs. 1 and 2, the cytoplasmic membrane is depicted as a single line and the internal membranous structures, which are connected with the cytoplasmic membrane and are constructed from membranes which resemble it when fixed in the corresponding manner, are also depicted as single lines. In other words, a 75-Å thick membrane of the "unit membrane" type is shown as a single line.

MEMBRANOUS STRUCTURES IN GRAM-NEGATIVE BACTERIA

In most gram-negative bacteria, the membranous structures are less well developed and simpler than those in gram-positive bacteria. The specialized cells of photosynthetic and chemosynthetic bacteria are an exception (Fig. 1). Simple membranous structures in the form of invaginations of the cytoplasmic membrane were first observed in *Spirillum serpens* and unidentified bacteria (Chapman and Kroll, 1957; Chapman, 1959). Differentiated lipid-rich regions were found

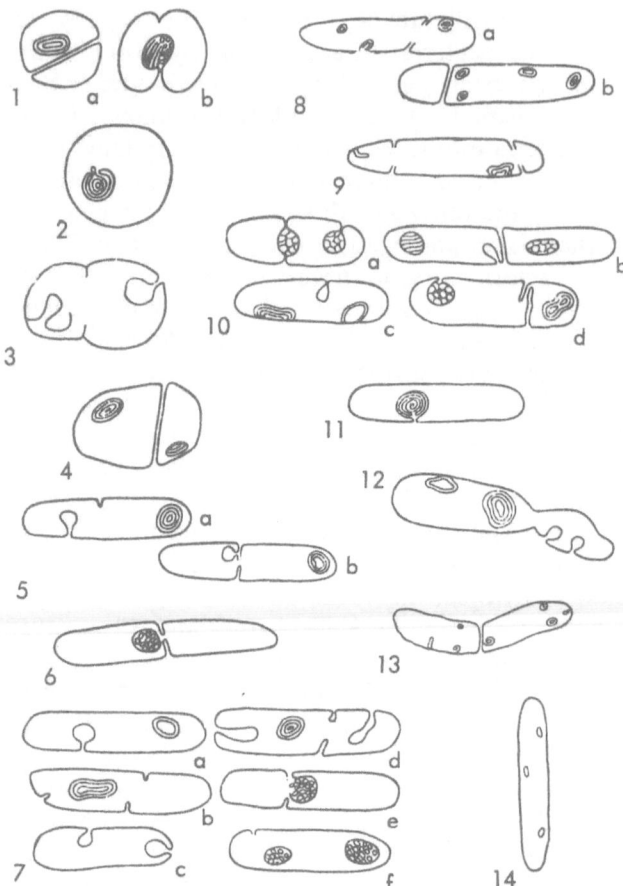

Fig. 2. Membranous structures (mesosomes) in gram-positive bacteria. 1. *Micrococcus lysodeikticus* (*a*—Salton and Chapman, 1962; *b*—Murray, 1963); 2. *Micrococcus roseus* (Murray, 1963); 3. *Diplococcus pneumoniae* (Thomasz *et al.,* 1964); 4. *Dermatophilus congolensis* (Gordon and Edwards, 1963); 5. *Bacillus megaterium* (*a*—Giesbrecht, 1960; *b*—Fitz-James, 1960); 6. *Bacillus coagulans* (Ohye and Murrell, 1962); 7. *Bacillus subtilis* (*a*—Van Iterson, 1961; *b*—Glauert *et al.,* 1961; *c*—Ryter and Jacob, 1963; *d*—Van Iterson and Leene, 1964*a; e*—Eiserling and Romig, 1962; *f*—Kawata, Inoue, and Takagi, 1963; 8. *Bacillus mycoides* (*a*—Tikhonenko and Bespalova, 1964; *b*—Malatyan and Biryuzova, 1965); 9. *Bacillus cereus* (Avakyan *et al.,* 1965); 10. *Listeria monocytogenes* (*a*—Grund, 1963; *b*—North, 1963; *c*—Kawata, 1963; *d*—Edwards and Stevens, 1963); 11. *Mycobacterium* sp. (Imaeda and Ogura, 1963); 12. *Corynebacterium diphtheriae* (Pavlova, 1964); 13. *Lactobacillus acidophilus* (Kodicek, 1963); 14. *Clostridium pectinovorum* (Fitz-James, 1962).

close to the wall in *E. coli* (Niklowitz, 1958). The fine structure of these regions was later revealed by Biryuzova (1960), who showed that the cytoplasm of *E. coli* contains groups of membranes situated on the periphery of the cell. In contrast to Niklowitz and Biryuzova, some authors could not find membranous structures in *E. coli* (Kellenberger and Ryter 1958; Conti and Gettner, 1962; Ogura, 1963; Sedar and Burde, 1965). Kran (1962), however, found looplike invaginations of the cytoplasmic membrane, and vesicular invaginations of the cytoplasmic membrane were later observed in *E. coli* cells after phage lysis (Cota-Robles and Coffman, 1964). Vanderwinkel and Murray (1962) found more complex structures—a system of membranes arranged in concentric layers on the periphery of the *E. coli* cell. According to the terminology proposed by Fitz-James (1960), these structures can be called mesosomes. Autolysis may cause these compact structures to straighten out and produce a row of membranes unconnected with one another, like those observed by Kushnarev and Pereverzev (1964). In the opinion of Van Iterson and Leene (1964*b*), however, the mesosomal structures in *E. coli* are merely a result of unfavorable growth conditions and normal *E. coli,* like most other gram-positive bacteria, do not have concentric layers of membranes. It is of interest that Edwards and Stevens (1963) could not find membranous structures in *E. coli* in a very thorough examination. These authors, however, attribute the appearance of membranous structures not to unfavorable growth conditions, but to a period of increased biosynthesis or enhanced exchange with the environment.

No membranous structures at all could be found in ultrathin sections of some representatives of gram-negative bacteria. Hence, a method of detecting them was devised, involving the introduction of a hydrogen acceptor into the cell, followed by fixation and the preparation of sections. Tetrazolium salts and tellurite are usually used. The membranous structures are demonstrated by the deposition of the reduced form of these compounds. The function of the membranous structures and the nature of the reduction reaction will be discussed later. We will confine ourselves here to an account of some of the information which this method has provided regarding the presence of membranous structures. Nermut (1963) and Nermut and Rýc (1964) found peripheral regions of cytoplasm where reduced tellurite or nitro-blue tetrazolium (formazan) was deposited and they regarded these regions as membranous structures. In contrast to these authors, Van Iterson and Leene (1964*b*) could not find in the cytoplasm of *Proteus vulgaris* any distinctly differentiated regions where the reduced indicators were deposited. Using tellurite they found only conglomerates of structural elements dispersed in the cytoplasm on the periphery of the cell. Without special treatment, the membranous formations in the cells of *S. serpens* appear as simple canals running from the cytoplasmic membrane into the cytoplasm (Murray, 1960; Murray and Birch-Andersen, 1963). The use of tetrazolium chloride revealed that *S. serpens* cells had not only invaginations of the cytoplasmic membrane in the form of separate tubules, but systems of membranes (mesosomes), situated on the periphery of the cell (Vanderwinkel and Murray, 1962).

Although the "tellurite" method of demonstrating membranous structures gives good results, one must agree with Kellenberger and Ryter's view (1964) that the fact that membranous structures cannot be directly detected in sections fixed with osmium or permanganate indicates that they are either structures of a special nature or, possibly, artifacts. The possibility that they are artifacts cannot be ruled out, since the cytoplasmic membrane, which can be taken as a standard, is quite clearly revealed without tellurite or tetrazole in fixation conditions where the membranous structures cannot be seen. Artifacts are also possible in the case where tetrazolium derivatives soluble in lipids are introduced into the cell (Weibull, 1953*a*; Takagi *et al.,* 1963). This was clearly shown in the experiments of Sedar and Burde (1965), who used 2,2′,5,5′-tetra-p-nitrophenyl-3,3′-(3,3′-dimethoxy-4,4′-biphenylene)-ditetrazolium chloride, which has no affinity for lipids. In 15- to 18-hr-old *E. coli* cells, formazan was deposited only in the cytoplasmic membrane and no intracellular aggregates which could be regarded as mesosomes were revealed. The use of nitro-blue tetrazolium brought out very distinctly the membranous structures in *Fusobacterium polymorphum.* The formazan was located in the cytoplasmic membrane and the mesosomes. Triphenyltetrazolium chloride did not give such a distinct picture (Takagi *et al.,* 1963). In view of this, attempts should be made with tellurite or appropriate tetrazolium derivatives to discover membranous structures in other bacteria where they are not revealed in sections. For instance, Brown *et al.* (1962) could not find any internal membranes in a marine pseudomonad. Such structures could not be found in the cells of *Alcaligenes faecalis* (Beer, 1960). Only shallow invaginations of the cytoplasmic membrane have been found in the cells of *Acetobacter suboxydans* (Claus and Roth, 1964).

The internal structure of the spirochaetes *Treponema microdentium* and *Borrelia recurrentis* is peculiar. Sections reveal several membranes arranged in concentric layers (Listgarten *et al.,* 1963; Ludýik, 1964).

The membranous systems of *Treponema* Reiter are distinctly seen, especially if the cell wall has been digested by trypsin (Kawata and Inoue, 1964). According to Ryter and Pillot (1963), however, the membranous systems of this *spirochaete* are poorly differentiated.

The appearance of membranous structures and their shape and position may depend on the stage of development of the bacteria. As was shown for *P. vulgaris* by Nikitina and Biryuzova (1965), Dart and Mercer (1963*a*) could not find any internal membranes in nodule bacilli. Conversion to the bacteroid form, corresponding to the formation of leghemoglobin, led to the appearance of internal membranes (Dart and Mercer, 1963*b*; Dixon, 1964). In this case the membranes are not densely packed and arranged in concentric layers. The membrane system consists of numerous vesicles, rarely lamellae, with anastomoses.

Some bacteria possess membranous structures of different complexity. For instance, *Neisseria gonorrhea* and *Brucella abortus* contain simple looplike invaginations of the cytoplasmic membrane and isolated membranous structures (mesosomes) in the center of the cell (Fitz-James, 1964*a*).

The membranous structures in *Azotobacter* cells have been the subject of several investigations. Small invaginations over the whole surface of the cytoplasmic membrane have been found (Wyss *et al.,* 1961). A membrane system in the form of canals penetrating the cytoplasm has also been discovered (Pangborn *et al.,* 1962; Van Iterson, 1963; Yakovlev and Levchenko, 1964). *Azotobacter* apparently possesses the most highly developed system of internal membranes among the gram-negative organisms. It is of interest that a well-developed membrane system is found in *Azotobacter* cysts (Tchan *et al.,* 1962). According to these authors, the cysts can be regarded as a vegetative cell invested with a very strong and thick wall.

Mesosomes are well developed in representatives of the Caulobacteraceae.They are located at the site of constriction of the cell (*Caulobacter bacteroides, Caulobacter crescentus, Caulobacter fusi-*

formis), in the region of the septum (*Asticcacaulis excentricus*), or at the base of the "tail," a feature peculiar to these cells (Stove, Poindexter and Cohen-Bazire, 1964).

Membranous structures are also present in such specialized bacteria as autotrophs—photosynthetic and chemosynthetic bacteria. The cells of *R. rubrum* contain small membranous structures, which have the appearance of vesicles and form a continuous system (Holt and Marr, 1965*a*, 1965*b*). They carry the pigments for photosynthesis and are called chromatophores. These structures are formed by invagination of the cytoplasmic membrane (Cohen-Bazire and Kunizawa, 1963; Boatman, 1964; Hickman and Frenkel, 1965*a*). The photosynthetic apparatus can have a different structure, as in *Rhodospirillum molischianum,* where it consists of stacks of parallel membranes, which are also connected with the cytoplasmic membrane (Giesbrecht and Drews, 1962; Hickman and Frenkel, 1965*b*). No differentiated chromatophores have been found in *Rhodomicrobium vanniellii.* Their function is performed by a system of concentric membranes located on the periphery of the cell under the cytoplasmic membrane (Boatman and Douglas, 1961). Similar membranous structures are found in *Hyphomicrobium sp.,* which lack chlorophyll but are apparently close "relatives" of *R. vanniellii* (Conti and Hirsch, 1965). Recently *Chlorobium* cells were found to contain mesosomal structures typical of gram-negative bacteria and chromatophores in the form of vesicles connected with the cytoplasmic membrane (Cohen-Bazire *et al.,* 1964). Some photosynthetic bacteria can become heterotrophic in darkness and take up oxygen. The photosynthetic apparatus is greatly reduced in this case, and only a few chromatophores, anatomically connected with the cytoplasmic membrane, are left (Cohen-Bazire and Kunizawa, 1963; Boatman, 1964). The respiratory chain in photosynthetic bacteria grown in darkness and in aerobic conditions is apparently not connected with the chromatophore system and is probably located in the cytoplasmic membrane and its associated invaginations.

Membranous systems are particularly well developed in another group of autotrophic bacteria, the chemoautotrophs (Murray, 1963; Murray and Watson, 1965). *Nitrosomonas europea* has a system of unit membranes (75 Å) arranged in four rows parallel to the cytoplasmic membrane. In *Nitrobacter agilis,* there are more unit membranes, arranged in pairs, thus forming double membranes. The whole cell of *Nitrosocystis oceanus* is penetrated by a system of 20–25 parallel membranes and recalls the grana of chloroplasts.

Thus, among the gram-negative bacteria there are aerobic forms (*E. coli, S. serpens, A. suboxydans, P. vulgaris*) in which the membranous structures are simple, consist of invaginations of the cytoplasmic membrane, and are sometimes very poorly developed. At the same time, in anaerobic spirochaetes, photosynthetic bacteria, and chemosynthetic bacteria, the internal membranes are very well developed and are characterized by considerable polymorphism. The membranous systems in *Caulobacter sp.* and *Hyphomicrobium sp.* are well developed.

MEMBRANOUS STRUCTURES IN GRAM-POSITIVE BACTERIA

The cells of gram-positive bacteria are characterized by a well-developed system of membranes. The membranous formations differ in size, structure, shape, and time of appearance in the cell cytoplasm. Fixation with permanganate or osmium tetroxide (OsO_4) is sufficient to reveal the membranous structures in sections of gram-positive forms of bacteria.

Different authors have given different names to the structures which they have found. The proposed names include mesosomes, chondrioids, peripheral bodies, plasmalemmosomes, and mitochondrial equivalents. In this discussion we will use the term mesosomes, since this term is most common in the literature and the functions of the membranous structures of bacteria are still far from clear. Although this term initially applied only to gram-positive bacteria, it was subsequently used to describe the membranous structures in gram-negative bacteria. The membranous structures of gram-positive bacteria usually have the form or organelles, although simple invaginations of the cytoplasmic membrane are also found. We will discuss first the structure of the membranous systems in spore-forming cells (Fig. 2).

Giesbrecht (1960, 1962) showed the presence of membranous structures in cells of *Bacillus megaterium.* They were situated in different regions of the cell: close to the wall, in the region of the nuclear material, at the site where the septum forms during division, and also in the spores. According to Giesbrecht, these membranous structures are formed by invagination of the cytoplasmic membrane and are identical with mitochondria, since they have a system of tubules formed by double membranes, which, in Giesbrecht's opinion, can be regarded as analogs of the mitochondrial cristae. However, in an investigation of the same bacteria and of *Bacillus medusa* and *Bacillus cereus* Fitz-James (1960) did not find any double membranes of the crista type. He found that the membranous structures result from invagination of the cytoplasmic membrane when the spores are formed.

Ohye and Murrell (1962) confirmed in the case of *Bacillus coagulans* that the mesosomes are associated with the initial stages of spore formation. Van Iterson (1961) shows electron micrographs of *Bacillus subtilis*. The mesosomes are found in the nuclear region and in the region adjoining the cytoplasmic membrane, particularly near the newly formed septum. They have the appearance of a system of concentric membranes or clusters of minute vacuoles. Such structures, formed by invagination of the cytoplasmic membrane, have been found by other authors in *B. subtilis* (Glauert *et al.,* 1961; Eiserling and Romig, 1962). Formazan and tellurite help reveal the mesosomes in *B. subtilis* (Vanderwinkel and Murray, 1962; Van Iterson and Leene, 1964*a*). Mesosomes in the form of densely packed membranes formed by invagination of the cytoplasmic membrane have been demonstrated in *B. subtilis* and *B. cereus* (Fig. 3) (Ryter and Jacob, 1963; Avakyan *et al.,* 1966).

Salton and Chapman (1962) have published interesting observations which help to clarify the structure of cocci. The cells of *Micrococcus lysodeikticus* have mesosomes about 2500 Å in diameter, composed of several membranes (up to five) arranged in concentric layers. These membranous structures originate in the cytoplasmic membrane and appear in the region of the septum (Murray, 1963). Two types of internal membranous structures differing from those of *M. lysodeikticus* have been found in *Diplococcus pneumoniae* (Tomacz *et al.,* 1964). The membranes of the first type are lamellar invaginations of the cytoplasmic membrane which precede the formation of the septum. These membranes are never folded in upon themselves at any point, but consist of two parallel single membranes. The second type of membranous structures—mesosomes—consists of a system of globules and vesicles which originate close to the transverse septum.

The cells of *Listeria monocytogenes* have a well-developed mem-

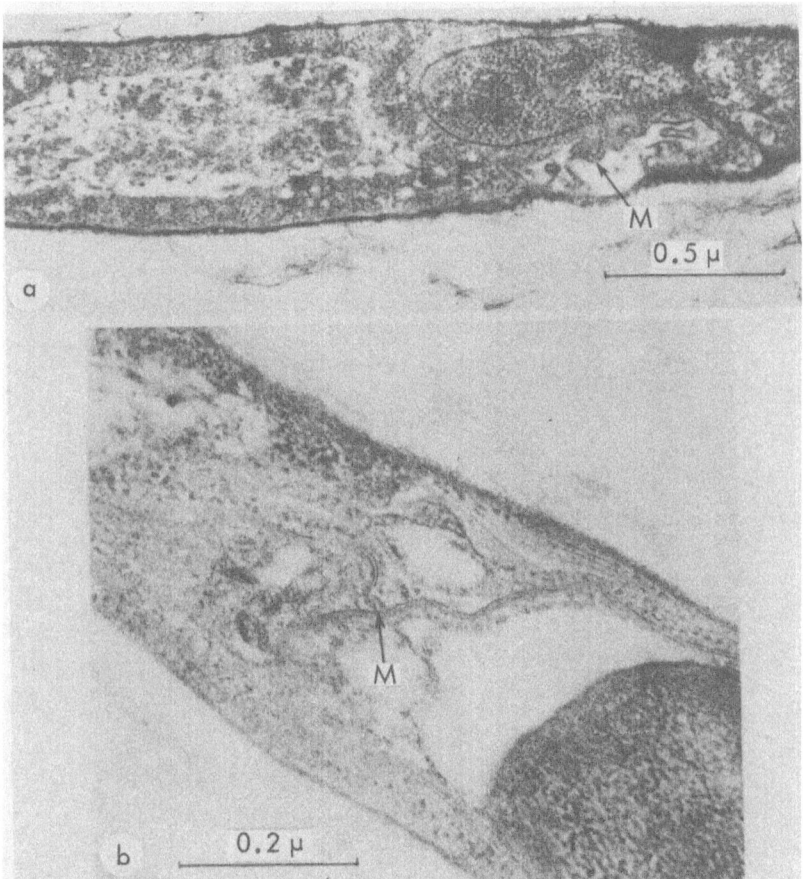

Fig. 3. *Bacillus cereus*—30-hr culture on meat-peptone agar; stage of spore formation —(M—membranous structures). (*a*) × 50,000; (*b*) × 100,000 (from Avakyan *et al.*, 1966).

brane system (Kawata, 1963; North, 1963; Grund, 1963; Edwards and Stevens, 1963). The mesosomes consist of lamellar and tubular structures of varying shape and size, located on the periphery of the cell and in the region of the nuclear substance.

Mesosomes are exceptionally well developed in various representatives of the mycobacteria. They were first described in *Mycobacterium tuberculosis* (Shinohara *et al.*, 1957). Koike and Takeya (1961) found that the mesosomes in the cells of *Mycobacterium lepraemurium* are built of several layers of membranes (80 Å thick), have no

limiting membrane, and are formed by invagination of the cytoplasmic membrane. Mesosomes which in section appeared like clusters
of vesicles were later found in addition to the layered mesosomes
(Imaeda and Convit, 1962; Imaeda and Ogura, 1963). Mesosomes in
the form of concentric layers of membranes (Fig. 4) connected with

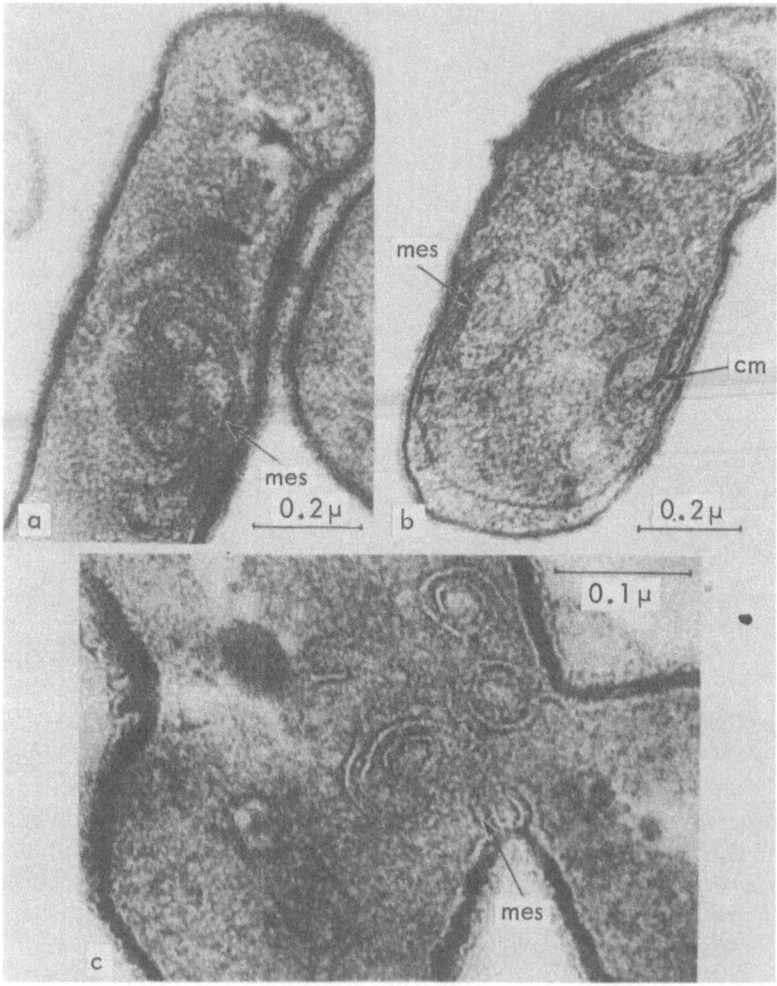

Fig. 4. *Corynebacterium diphtheriae* — seven-day culture — period of maximum
toxin production — (cm — cytoplasmic membrane; mes — mesosome). (*a*) ×
70,000; (*b*) × 70,000; (*c*) × 180,000 (from Pavlova, 1964).

the cytoplasmic membrane have been observed in diphtheria bacteria (Pavlova, 1964; Barban, 1963).

A comparison of the information summed up in Figs. 1 and 2 shows that the most diverse bacteria (aerobes, facultative and obligate anaerobes, photosynthetic bacteria) contain membranous structures that are polymorphic not only in a group of bacteria, but even in the same bacterium.

Internal membranes have been found even in the smallest bacteria, the *Mycoplasma*. It is true that it is difficult at present to decide if they are the initial stage of development of the elementary bodies (Domermuth *et al.,* 1964*a, b*). Even the rickettsiae, which come between bacteria and viruses, have looplike invaginations of the cytoplasmic membrane. This is a feature common to them and gram-negative bacteria (Ito and Vinson, 1965).

The variety of forms of internal membranous structures in bacteria (simple invaginations of the cytoplasmic membrane in the form of loops or vesicles, membranes lying parallel to the cytoplasmic membrane, chromatophores, and mesosomes of vacuolar, lamellar, or tubular structure) indicates the variable, evolutionally unestablished form of these structures of the bacterial cell. This is one of the main differences between the membranous structures of bacteria and mitochondria. The membranes of bacteria do not resemble an endoplasmic reticulum. They occupy a relatively small part of the cytoplasm, and only in *Azotobacter* do they form a membranous network which penetrates the cell. It is obvious that the structure of the membranous systems observed in ultrathin sections depends to some extent on the direction of the cut relative to the surface of the organelle (Kellenberger and Ryter, 1964).

INTERRELATIONSHIP OF INTERNAL MEMBRANES AND CYTOPLASMIC MEMBRANE

Despite the fact that the mesosomes sometimes occupy a considerable part of the bacterial cell, they are not independent structures. The mesosomes are derived from the cytoplasmic membrane and remain anatomically connected with it. This has been demonstrated by almost all the workers cited above. The connection of the mesosomes with the cytoplasmic membrane has been indirectly proved by experiments on *B. megaterium* and *B. subtilis,* where it has been shown that the mesosome is degraded to a considerable extent by stretching when bacterial protoplasts are obtained. This degradation is obviously due to redistribution of the mechanical stresses in the cytoplasmic membrane (Fitz-James, 1964*b*, *c*; Ryter and Landman, 1964). It is of interest that the chromatophores of photosynthetic bacteria are also formed from the cytoplasmic membrane (Kondrat'eva, 1963, 1964; Giesbrecht and Drews, 1962; Hickman and Frenkel 1965*a*, *b*).

In size and shape, the internal membranous structures resemble the cytoplasmic membrane. According to different authors (see Figs. 1 and 2), the internal membranes, like the cytoplasmic membrane, are no more than 70–90 Å thick; i.e., these are single membranes ("unit membranes," according to Robertson).

Thus, all the membranous structures of bacteria, from the most primitive to the most highly developed—the mesosomes and chromatophores—are derived from the cytoplasmic membrane.

ISOLATION OF BACTERIAL MEMBRANES AND THEIR STRUCTURE

There are several methods permitting mild destruction of the bacterial cell for the isolation of the membranous structures without their disintegration into small fragments. The commonest methods include enzymatic digestion of the cell wall with lysozyme (Weibull, 1956). Phage lysis is sometimes used (Cota-Robles and Coffmann, 1964). The enzymatic lysis of bacterial cells and other methods of disrupting them to obtain cell fragments have been discussed in several review articles (De Ley, 1963; Hughes and Cunningham, 1963; Salton, 1964; Stolp and Starr, 1965).

The action of lysozyme on gram-positive bacteria completely dissolves the cell wall. In an isotonic medium (0.8–$1\ M$ sucrose) the protoplast remains intact and becomes rounded, but a sharp reduction in osmotic strength leads to shock and the release of the cytoplasmic membrane and membranous elements of the cell. The cytoplasmic membrane was first isolated from *B. megaterium* by Weibull (1953*b*). The membranous structures are usually isolated with the cytoplasmic membrane. This confirms their anatomical connection in the cell.

Protoplasts can also be lysed by "metabolic lysis" (Wachsman and Storck, 1960; Abrams, 1959). "Metabolic lysis" of the protoplasts of *B. megaterium* and *Streptococcus faecalis* is due to accumulation in

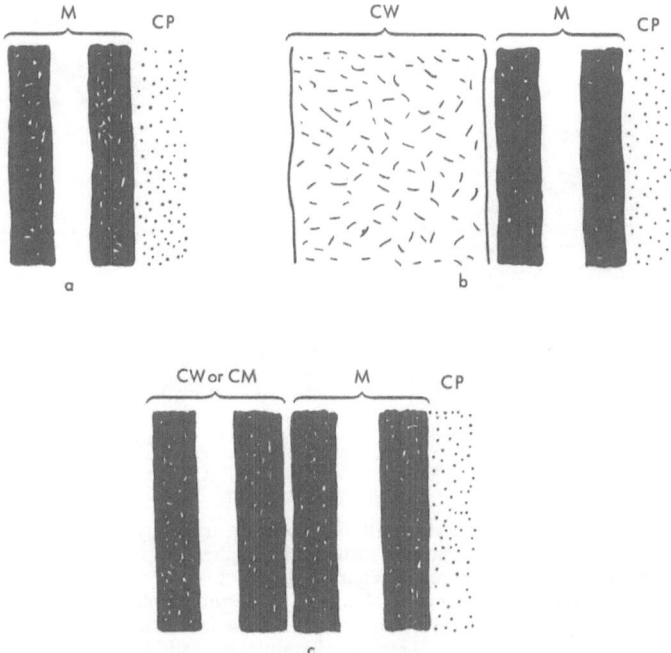

Fig. 5. Main types of surface structures in bacteria (CP stands for cytoplasm): (a) Single "unit membrane" M found at surface of some bacteria
and protoplasts. (b) Thick amorphous cell wall CW and cytoplasmic
membrane M, as in many gram-positive bacteria. (c) Multilayered "wall"
or compound membrane CM and cytoplasmic membrane M, as in gram-
negative bacteria (from Salton, 1964).

the cytoplasm of large amounts of soluble, low-molecular metabolic
products formed by the oxidation of propionate or lactate.

In the case of gram-negative bacteria, which have a cell wall of
different composition and structure from that of gram-positive bacteria (Fig. 5), the action of lysozyme in the presence of EDTA (Repaske, 1958) leads to the formation of "spheroplasts." Spheroplasts
retain two membranes—a membrane which forms part of the cell
wall and the underlying cytoplasmic membrane (Salton, 1964). The
external membrane of the cell wall may not extend over the whole
surface of the spheroplast (Fig. 6). Like protoplasts, spheroplasts
undergo osmotic shock if placed in a hypotonic medium.

When the concentration of osmotically active substances (salts,
sucrose, polyethylene glycol) in the medium is greatly reduced and
water penetrates rapidly into the protoplast, the "packing" of the

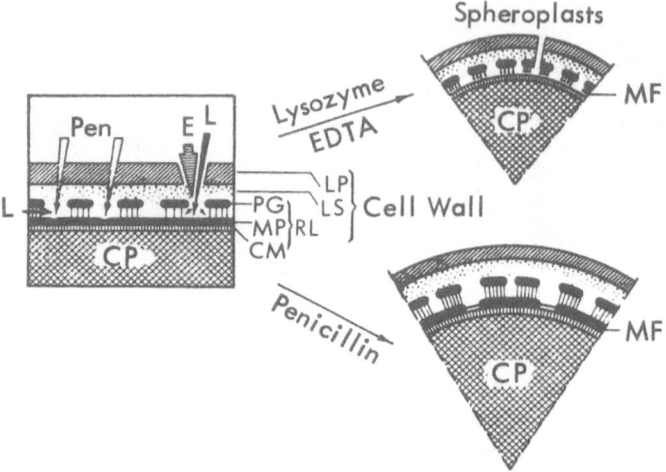

Fig. 6. Diagrammatic representation of complex, triple-layered cell wall of gram-negative bacteria and formation of spheroplasts. LP — lipoprotein layer; LS — lipopolysaccharide layer; RL — rigid layer; PG — protein granules; MP — mucopolymer; CM — cytoplasmic membrane; CP — cytoplasm; MF — mucopolymer fragments; Pen — penicillin; L — lysozyme; E, EDTA — ethylenediaminetetraacetic acid; P — protease. Lysozyme L, aided by EDTA and penicillin, induced depolymerization of the mucopolymer in the rigid layer. After lysozyme treatment the cell wall retains small mucopolymer fragments MF. Penicillin spheroplasts and related L-forms may retain large mucopolymer fragments or an entire modified nonrigid mucopolymer.

membranous structures in the cell may be altered. Hence, isolated membranous structures differ from the membranous structures observed in the intact cell. When the protoplasts of some species of bacilli are formed, the mesosomes are greatly degraded (Fitz-James, 1964; Ryter and Landman, 1964; Ryter and Jacob, 1963). Osmotic shock of the protoplast leads to even greater changes in mesosome structure. For instance, ultrathin sections of the membranous fraction of *M. lysodeikticus* (Fig. 7) show that in addition to the cytoplasmic membrane (80–100 Å) there is a system of internal membranes of the same thickness (Lukoyanova, Gel'man and Biryuzova, 1961). In intact *M. lysodeikticus* cells, Salton and Chapman found in the center of the cell membranous structures in the form of closely packed concentric layers of single membranes (75–80 Å) identical in thickness and structure to the cytoplasmic membrane (see Fig. 2). Lysis of the cell evidently leads to straightening and stretching of

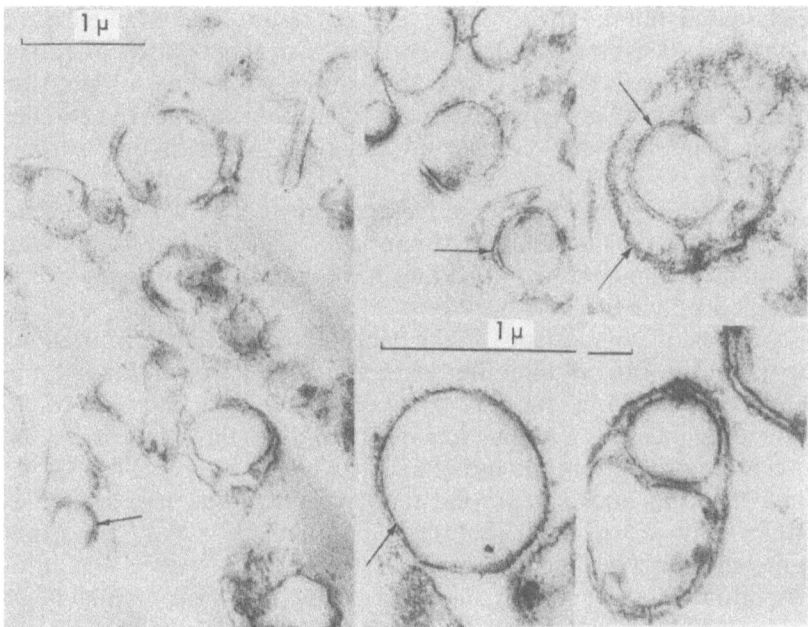

Fig. 7. Isolated membranous system of *Micrococcus lysodeikticus.*

the membranes, producing the picture which we observed—a system of internal membranes (up to five) freely arranged inside the cytoplasmic membrane.

The membranous structures isolated from *Azotobacter agilis* have a different structure (Pangborn *et al.,* 1962; Robrish and Marr, 1962). There is no system of concentric membranes, but there is a very complex system of tubules. The authors observed practically the same picture in intact cells (see Fig. 1). When the membranous structures are arranged diffusely in the cell, their organization is presumably only altered slightly by osmotic shock of the cell.

The chromatophore apparatus, which consists of numerous invaginations of the cytoplasmic membrane (see Fig. 1), does not undergo any great changes when isolated from the cell. The chromatophore system retains its shape and connection with the cytoplasmic membrane (Boatman, 1964). Isolated chromatophores, which have the form of elongated vesicles, are similar to the chromatophores in the intact cell (Cohen-Bazire *et al.,* 1964). The stacks of lamellae in *R. molischianum* become less densely packed when the cell is lysed, but the shape and thickness of the individual lamellae are unaltered (Giesbrecht and Drews, 1962). Chromatophores obtained from *Rhodopseudomonas spheroides* differ with their origin: the "heavy fraction" is derived from the cytoplasmic membrane and the "light fraction" from vesicular invaginations (Worden and Sistrom, 1964).

Sections of membranes of lactic acid bacteria (*S. faecalis*) were obtained for the first time by Abrams *et al.* (1964). In addition to the cytoplasmic membrane (80 Å) they have internal membranes—vesicles and tubules.

From a comparison of the morphology of the membranous structures in the cell and after extraction, we can postulate that the change in "packing" on lysis of the cell is due to the absence of the bounding membrane, which imparts strength to the whole organelle. According to Salton, the use of agents which inhibit swelling of the mitochondria does not prevent the "expansion" of the mesosomes when they are isolated (Salton, 1964). This is a very important difference between the membranous structures of bacteria and mitochondria. The latter have a semipermeable membrane which helps the mitochondria to retain their shape when they are isolated and performs several functions (Chappell and Greville, 1963). It has been suggested that the resistance of bacteria to freezing and thawing is due to the fact that they do not have the membrane-bound organelles typical of higher organisms (Hechter, 1965).

FUNCTIONS OF CYTOPLASMIC MEMBRANE AND MEMBRANOUS STRUCTURES

Although the bacterial cell is very small it can accomplish the main metabolic processes—respiration, absorption and excretion of substances, and the biosynthesis of proteins, nucleic acids, and other compounds. All these processes must be coordinated with one another and spatially organized in the small bacterial cell. In highly organized cells, protein biosynthesis is associated with the ribosomes and the nucleus, the respiratory chain of enzymes and the Krebs cycle is associated with the mitochondria, and the penetration of substances into the cell, with the plasma membrane.

In bacteria the question of protein synthesis is resolved relatively simply, since they have a nuclear substance and polysomes. However, as regards structural elements constructed from lipoproteins, the bacterial cell occupies a special position. As the preceding account has shown, bacteria have no mitochondria, but they do have a cytoplasmic membrane and membranous structures anatomically linked with it. The structures can differ considerably not only in different bacteria but also in one bacterium at different stages of development. In what structural elements is the respiratory chain located—the cytoplasmic membrane or the membranous structures? Do the bacterial membranes perform any other specific functions? An exhaustive answer to these two questions is impossible at present, although there is abundant information regarding the role of the membranous structures. However, paradoxical as it may appear, research workers are dubious about the role of the membranous structures in bacterial metabolism, not so much because they cannot discover any functions for them, as because too many functions have been found for them (Kodicek, 1963).

Cell division and sporulation can be included among the specific processes which involve the cytoplasmic membrane and the mesosomes (Fitz-James, 1965; Cole, 1965). In gram-negative bacteria, where the mesosomes are poorly developed or absent altogether, as in *E. coli, S. serpens, N. gonorrheae,* and *Fusobacterium,* invaginations of the cytoplasmic membrane appear during cell division (see Fig. 1). In *Thiovolum majus,* invagination of the lipoprotein membrane of the wall and the underlying cytoplasmic membrane occurs between two large lamellar structures similar to the mesosomes of gram-positive bacteria (De Boer *et al.,* 1961). Some authors indicate that the mesosomes are implicated in division and sporulation in gram-positive bacteria. Imaeda and Ogura (1963) think that the mesosomes of mycobacteria produce the material of the new cell wall during division.

An investigation of the synthesis of lipoamino acids in sarcinae, staphylococci, and lactobacilli showed that the formation of amino-acid esters and phosphatidylglycerol, which are components of the membranes, is closely connected with the phase of growth, reaching a maximum in the exponential phase and declining in the stationary phase (Macfarlane, 1964*a, b*). It is possible that in dividing cells a new cytoplasmic membrane is synthesized by the formation of mesosomes. Ryter and Landman (1964) found a connection between the mesosomes and the synthesis of the cell wall in L-forms of *B. subtilis.*

Kawata (1963) observed the process of cell division in *L. monocytogenes.* The cytoplasmic membrane was centripetally invaginated into the cytoplasm and subsequently gave rise to mesosomes and a new cell wall.

Ohye and Murrell (1962) investigated sporulation in *B. coagulans* and observed that the membrane of the future spore was formed by invagination of the cytoplasmic membrane. These authors point out the importance of the mesosomes for the formation of septal partitions and the development of spore membranes. The importance of the mesosomes for cell division and sporulation in gram-negative bacteria was noted by Van Iterson (1961), Kawata *et al.* (1963), and Ryter (1965) for *B. subtilis,* by Fitz-James (1960) for *B. megaterium,* by Glauert and Hopwood (1960) for *Streptomyces coelicolor,* and by other authors. It is significant that mesosomes are involved in sporulation in aerobes and in obligate anaerobes, as Fitz-James (1962) showed for butyric acid bacteria.

There is interesting information indicating a connection between the mesosomes and the functions of the nucleus. Ryter and Jacob (1963) in a study of the interrelationships of the mesosomes and nucleus in *B. subtilis* established the existence of a connection between the nucleus and cytoplasmic membrane through one or two mesosomes. They showed that mesosomes are formed, shift about, and may disappear during the cycle of cell division, i.e., that these structures are not always present in the bacterial cell. Further investigations showed that the mesosome of *B. subtilis* is connected so strongly with DNA that it is extracted with it during lysis. Replication of the bacterial chromosomes and their distribution in the daughter cells involve the mesosomes, which are specific orienters of this process (Jacob *et al.,* 1966). Lark (1966) also discusses the role of the cytoplasmic membrane in chromosome replication in *E. coli.* Van Iterson believes that the connection of the mesosomes with the nuclear material in *B. subtilis* may be very significant: the mesosomes may originate or develop among the threads of nuclear material (Van Iterson, 1961). Giesbrecht (1962) observed that the nuclear substance was connected with the membranous structures and, through them, with the cell surface in *B. megaterium.* Recently, biochemical evidence of the connection of the DNA in *B. subtilis* with the membranous system has been obtained (Ganesan and Lederberg, 1955). According to Fuchs (1965), there is a connection between the position of the mesosomes in the cell and the division cycle, but there is no reliable information in support of the participation of mesosomes in the division of the nuclear substance. Malatyan's work (1963) on

E. coli showed that during the growth of the culture the number of mesosomes increased in parallel with cell division. At the same time, Dart and Mercer (1963*b*) found no connection between the membranous system of nodule bacteria (rhizobium) and multiplication.

The "trivial" functions associated with the lipoprotein membranes of highly organized cells include the attachment of the ribosomes (endoplasmic reticulum), organization of the respiratory chain enzymes (mitochondria), absorption of substances, and production of metabolites. These functions are also found in bacterial membranes.

From morphological information, Glauert (1962) thinks it unlikely that the membranous structures of bacteria perform the role of an endoplasmic reticulum. However, protein synthesis is more rapid in ribosomes which retain their connection with the cytoplasmic membrane and internal membranes in preparations obtained from *S. faecalis* cells (Abrams *et al.,* 1964). Similar observations have been made on preparations of *E. coli, B. megaterium, Pseudomonas sp., Bacterium anitratum* (Tani and Hendler, 1964; Schlessinger, 1963; Butler *et al.,* 1958; Mitsui, 1961; Hallberg and Hauge, 1965; Chaloupka and Vereš, 1961). It appears that the apparatus for the synthesis of ribosomal RNA is also connected with the membranes (Nielsen and Abrams, 1964; Yudkin and Davies, 1965). Biochemical data indicating a connection between the polysomes and the membrane have been confirmed by Pfister and Lundgren (1964), who demonstrated that at times there is a connection between the polysomes and the cytoplasmic membrane or internal membranes of *B. cereus.* Aronson reports that the membranous apparatus of *B. megaterium* retains the ribosomes by means of a polypeptide formed in them (Aronson, 1966).

According to contemporary opinion, the products of protein biosynthesis are released via the endoplasmic reticulum. It is probable that, in the bacterial cell, connection of the polysomes with the lipoprotein membrane is a physiological necessity during active protein synthesis.

Other biosynthetic processes have been discovered in membrane preparations which obviously contained the cytoplasmic membrane and internal membranes. These processes included the synthesis of polysaccharides (Markovitz and Dorfman, 1962), phospholipids (Hill, 1962), and polyhydroxybutyric acid (Merrick and Doudoroff, 1961). Such membrane preparations from *S. faecalis, Streptococcus pyogenes,* and *B. megaterium* were found to contain enzymes which cleave ATP (Abrams *et al.,* 1960; Sokawa, 1965; Weibull *et al.,* 1962). According to Pavlova (1964), membranous structures are particularly well developed in diphtheria bacteria during the period of intense toxin formation. Nitrogen fixation in azotobacter cells is apparently associated with the internal membranes or their fragments (Yakovlev and Levchenko, 1964; Bulen *et al.,* 1964). According to Sokawa and Egami (1965), an exoenzyme of the streptolysin type is localized in the membrane of *S. pyogenes* and is released from the membrane by ATPase and magnesium ions.

In discussing the functions of the cytoplasmic membrane and the membranous structures in bacteria, we must not overlook the cells of autotrophs. In photosynthetic bacteria, the chromatophores are anatomically connected with the membrane, and sometimes photosynthesis is effected in the cytoplasmic membrane itself. The membranous apparatus is most highly developed in chemosynthetic bacteria. This is possibly due to the need to obtain energy from an "unsuitable" substrate, where there must be a large surface of contact of the enzymes with the substance undergoing oxidation.

The roles of the cytoplasmic membrane and internal membranes in the organization of the enzymes of the respiratory chain and oxidative phosphorylation are dealt with in special chapters; hence, we will mention only certain aspects here.

Numerous cytochemical observations indicate that oxidation—reduction indicators and fluorochromes are localized in the cytoplasmic membrane and the membranous structures. In addition to the cytochemical data there are also data indicating the presence of the respiratory chain enzymes in isolated membranous structures and the cytoplasmic membrane. These data are evidence in support of the view that the bacterial membrane system (cytoplasmic membrane and membranous structures) perform the functions of mitochondria. Some authors even identify the membranous structures with mitochondria. This would be justified if mesosomes were present in all aerobic bacteria and were absent in obligate and facultative anaerobes. One could then postulate that the amount of respiratory chain enzymes and, hence, energy production increased as a result of increase in the membrane surface by mesosome formation, thus satisfying the energy requirements for division, sporulation, nitrogen fixation, and other processes.

Most of the investigations dealing with the formation of mesosomes and their structure have been carried out on aerobic bacteria. Demonstration of these structures in facultative and obligate anaerobes is obviously of considerable interest for an understanding of their function. Vanderwinkel and Murray observed a reduction in the number of membranous structures in anaerobically grown *S. serpens.* However, when *M. lysodeikticus* is grown in anaerobic conditions, the mesosomes show no change (Murray, 1963). According to Edwards and Stevens, anaerobically grown cells of *L. monocytogenes* are similar to aerobes in the arrangement of mesosomes in the cell and in their number and structure. The shape and structure of mesosomes in *Myxococcus xanthus* cells are variable. At some stages of development of an aerobic culture the mesosomes may disappear, and they can be induced to appear by anaerobiosis (Voelz, 1965). In *Caulobacter sp.,* under poorly aerated conditions, the number of mesosomes and the cytochrome content of the cell are increased (Cohen-Bazire *et al.,* 1966).

Mesosomes are particularly well developed in the anaerobic *Lactobacillus acidophilus* (Glauert, 1962) and *Clostridium botulinum* (Takazi, *et al.,* 1965). Yet there is also information which indicates the absence of mesosomes in the anaerobic bacterium *Lactobacillus bifidus* (Overman and Pine, 1963). Obligate anaerobes (*Clostridium pectinovorum*) have mesosomes which appear to be connected with sporulation (Fitz-James, 1962). Yet in the nonsporulating anaerobic bacterium *Bacteroides ruminicola* the membranous structures are well developed (Bladen and Waters, 1963). Membranous structures are distinctly differentiated in cells of other anaerobic bacteria (*F. polymorphum*) and various spirochaetes (see Fig. 1).

Thus, no correlation has been found between the degree of aerobicity of the cell and the development of membranous structures in it. These structures are found in obligate aerobes, which have a complete respiratory chain similar to the mitochondrial chain, and in obligate anaerobes, which lack a respiratory chain and draw their energy from fermentation processes. In anaerobic bacteria, oxidation—reduction enzymes are not connected with the structural elements of the cell at all but pass into the soluble part when it is disintegrated (Hughes, 1962; Oparin *et al.,* 1964).

The membranous structures can apparently perform various functions, one of which is the organization of the enzymes of the respiratory chain and oxidative phosphorylation. Mitochondria and membranous structures, including the most fully developed—the mesosomes—have features of similarity and difference, the discovery of which is of great interest and can help to explain the development of such a complex structure as the mitochondrion in the process of evolution.

Permeability—the passage of substances into the bacterial cell— is associated solely with the cytoplasmic membrane (Mitchell, 1959*a*, 1963; Marquis and Gerhardt, 1964), although the participation of the internal membranes is not out of the question, since they are anatomically connected with the cytoplasmic membrane.

Thus, bacterial membranous systems are involved in various functions: division, sporulation, exchange with the ambient medium, biosynthesis, the respiratory chain, nitrogen fixation, photosynthesis, and chemosynthesis. As distinct from mitochondria, the membranous systems of bacteria contain, in addition to the respiratory chain and the mechanism for regulation of permeability, several other enzyme systems which are quite specific to bacteria. The polymorphism of the membranous structures in bacteria may be a manifestation of the diversity of function of these structures. In different bacteria the membranous structures may perform different functions. However, the hypothesis of polyfunctional bacterial membranes, i.e., that there are several enzyme systems located in these membranes, is equally probable. An attempt to relate two functions to the same part of the membrane system was made by Fitz-James, who takes a broad view of the functions of mesosomes: they play a special role in the formation of spores and of cell septa during division, and their main function is to supply energy to points of intensive growth. The mesosomes, like the cytoplasmic membrane with which they are connected, are the sites of localization of the enzyme systems—the energy suppliers, and in this sense they do perform the same function as mitochondria (Fitz-James, 1960). Similar attempts to "reconcile" the functions have been made in several investigations, but they will not be convincing until the methods of preliminary fractionation are improved. It is very difficult to understand how the spatial organization of such diverse processes in a continuous system (cytoplasmic membrane—invagination—mesosome) is achieved. The explanation should probably be sought in the polyfunctionality of the membranes, where there is an alternation of regions—loci of particular enzyme systems. Not only are we unable to differentiate preparatively or cytochemically the individual loci of enzymes in the membrane, but we cannot even separate such relatively coarse structures as the cytoplasmic-membrane and internal-membrane systems. Hence, we naturally cannot analyze the spatial organization of the various enzyme systems connected with the lipoprotein membranous components of the bacterial cell. In attempting to interpret the role of membranous structures in bacteria, we must not overlook the poor development of these structures in gram-negative bacteria nor the fact that there are gram-positive bacteria that lack internal membranes—*Micrococcus radiodurans* (Murray, 1963), for instance. Murray is probably right in his view that the membranous structures are not an essential element of bacterial structure but are a means of intensifying functions of the cytoplasmic membrane when necessary.

Chapter II

Molecular Organization of Bacterial Membranes

The most important processes of energy conversion (photosynthesis, chemosynthesis, and respiration) in bacteria are associated with the lipoprotein membranes, but bacteria do not have specialized organelles—(mitochondria or chloroplasts). The system of membranes in aerobic and most facultative anaerobic bacteria can be regarded as the functional analog of mitochondria. However, mitochondrial functions are more limited than those of the bacterial membranes. As discussed in the previous chapter, bacterial membranes are involved in a number of specific processes—division, sporulation, cell-wall formation, biosynthesis of proteins and other polymers, nitrogen fixation, and so on. Although none of these processes are associated with mitochondria, they have at least one feature in common—they require energy, which is produced in adjacent parts of the membranes.

At present there is no experimental method of differentiating regions of bacterial membranes according to the functions performed. So far it has been impossible even to separate the cytoplasmic membrane from the internal membranous structures, since they are anatomically connected and are extracted together. Hence, bacterial biochemistry at present is concerned with bacterial membranes as a whole.

Biological membranes are constructed of proteins and lipids and, hence, their structure and stability are determined by the nature, content, surface properties, and electrical charges of these components. Investigation of the principles of bacterial-membrane organization, i.e., the composition of the membranes and the way in which the protein and lipid components interact, is one possible approach towards an understanding of the specific structure of bacterial membranous structures and their differences from mitochondria.

33

Interest in the study of bacterial lipids, particularly membrane lipids, has recently increased. Information regarding the structural proteins involved in the formation of membranes is extremely limited. There is much more information relating to the properties of enzymatic proteins and their interaction with lipids. These questions receive special treatment in the appropriate sections of Chapter III.

LIPIDS OF BACTERIAL MEMBRANES

Since there have been few investigations of the lipids of bacterial membranes, it is advisable to start off with a very brief account of the most important features of the lipid composition of intact bacterial cells. An examination of the abundant material on bacterial lipids is beyond the scope of this book. We refer those interested to the recent excellent reviews (Asselineau and Lederer, 1960; Asselineau, 1962; O'Leary, 1962; Kates, 1964; Macfarlane, 1964a).

According to the general view, bacterial lipids differ greatly from the lipids of higher organisms in their high content of free fatty acids and the absence (usually) of sterols and phosphatidylcholine. In many families the phospholipids are poor in nitrogen and rich in carbohydrates. With some reserve the data for the majority of gram-positive bacteria can be taken to refer to their membranes, since up to 80% of all the cell lipids are localized in the cytoplasmic membrane and membranous structures of these bacteria, while the cell wall contains a very small amount or none at all. In gram-negative bacteria, whose cell wall is rich in lipids, the situation is not so simple.

The most regular constituents of bacterial phospholipids are acid phospholipids—mono- and diphosphatidylglycerols. Phosphatidylethanolamine (cephalin) and phosphatidylserine are found in some bacteria (Kennedy, 1964).

$$
\begin{array}{lcl}
& & \overset{\displaystyle O}{\overset{\|}{}} \\
CH_2OOCR & CH_2O-P-OCH_2 & \\
| & |\quad\ \ | & | \\
CHOOCR'\ O & CHOH\ \ OH\ \ CHOOCR'' & \\
|\qquad\ \ \| & | \\
CH_2O\!-\!\!-\!P\!-\!\!-OCH_2 & CH_2OOCR''' \\
\qquad\quad | \\
\qquad\quad OH
\end{array}
$$

Diphosphatidylglycerol

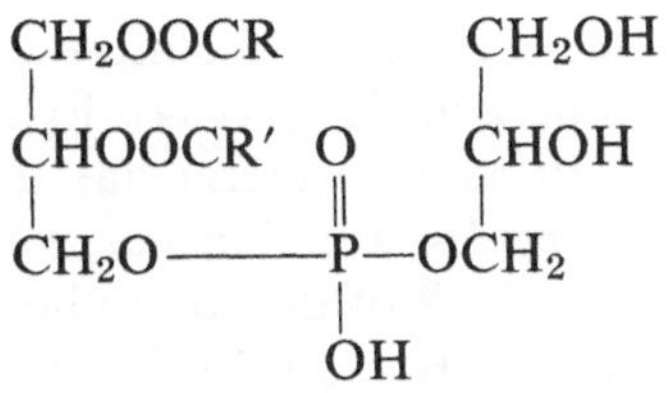

$$
\begin{array}{ll}
\text{CH}_2\text{OOCR} & \text{CH}_2\text{OH} \\
| & | \\
\text{CHOOCR}' \quad \text{O} & \text{CHOH} \\
| \quad\quad || & | \\
\text{CH}_2\text{O}\text{---}\text{P}\text{---}\text{OCH}_2 \\
\quad\quad\quad | \\
\quad\quad\quad \text{OH}
\end{array}
$$

Phosphatidylglycerol

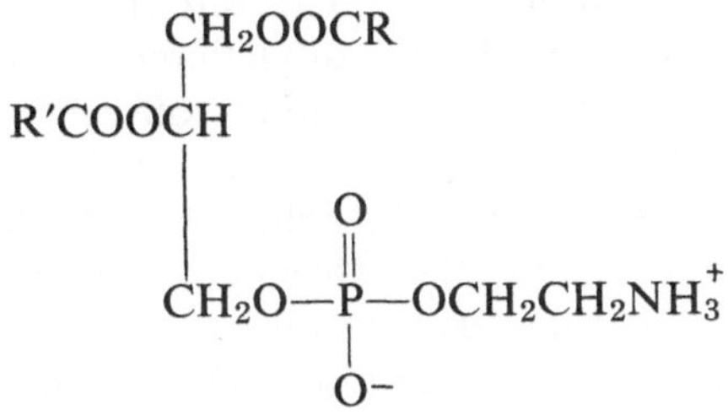

$$
\begin{array}{l}
\quad\quad \text{CH}_2\text{OOCR} \\
\quad\quad\quad | \\
\text{R}'\text{COOCH} \\
\quad\quad\quad | \\
\quad\quad\quad\quad\quad \text{O} \\
\quad\quad\quad\quad\quad || \\
\quad\quad \text{CH}_2\text{O}\text{---}\text{P}\text{---}\text{OCH}_2\text{CH}_2\text{NH}_3^+ \\
\quad\quad\quad\quad\quad | \\
\quad\quad\quad\quad\quad \text{O}^-
\end{array}
$$

Phosphatidylethanolamine

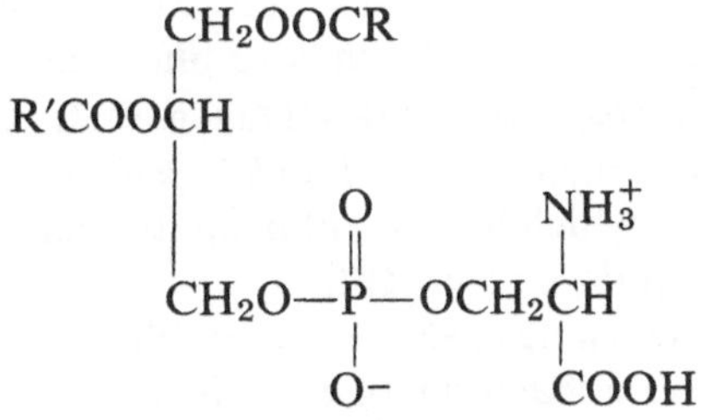

$$
\begin{array}{l}
\quad\quad \text{CH}_2\text{OOCR} \\
\quad\quad\quad | \\
\text{R}'\text{COOCH} \\
\quad\quad\quad | \\
\quad\quad\quad\quad\quad \text{O} \quad\quad \text{NH}_3^+ \\
\quad\quad\quad\quad\quad || \quad\quad | \\
\quad\quad \text{CH}_2\text{O}\text{---}\text{P}\text{---}\text{OCH}_2\text{CH} \\
\quad\quad\quad\quad\quad | \quad\quad\quad | \\
\quad\quad\quad\quad\quad \text{O}^- \quad\quad \text{COOH}
\end{array}
$$

Phosphatidylserine

Phosphatidycholine (lecithin) is not found in most of the investigated bacteria. Bases of the sphingosine type are not found either (Asselineau and Lederer, 1960). With a few exceptions (*Agrobacterium radiobacter* and *Agrobacterium rhizogenes*) bacteria lack enzyme systems for the synthesis of choline—a component of phosphatidylcholine (Goldfine and Ellis, 1964). Evidence that phosphatidylcholine is present in some bacteria has recently appeared. In *Rhodopseudomonas spheroides* and *Thiobacillus thiooxidans,* phosphatidylcholine makes up 20% of the total phospholipids (Lascelles and Szilagii, 1965; Jones and Benson, 1965). In cells of Hyphomicrobium NQ 521 up to 30% of the phospholipid phosphorus belongs to phosphatidylcholine (Hagen *et al.,* 1966). Thorne (1964) recently reported that the phospholipids of *Lactobacillus casei* contain up to 45% phosphatidylcholine, but this result does not agree with that of Ikawa (1963), who found no choline in this bacterium. As an illustration of the variation in the composition of bacterial phospholipids we can cite data on phosphatidylethanolamine content. In *B. cereus* cells, 50% of the phospholipids consists of phosphatidylethanolamine; in *Pseudomonas aeruginosa* and *E. coli* cells, the figure is 90 %, in *Salmonella typhimurium* cells, 84%, and in *T. thiooxidans* cells, 3% (Houtsmuller and Van Deenen, 1963; Singha and Gabi, 1964; Kaneshiro and Marr, 1962; Macfarlane, 1962a; Jones and Benson, 1965). Although the lipid composition of Hyphomicrobium NQ 521, *N. oceanus,* and *N. europea* cells all are very similar to one another as regards abundance of membranous structures, their lipid composition is different. For instance, in Hyphomicrobium 22% of the phospholipid phosphorus belongs to phosphatidylethanolamine, 36% to phosphatidyl-N-dimethylethanolamine, 30% to phosphatidylcholine, and 10% to polyglycerophosphatides, in *N. oceanus* 67% belongs to phosphatidylethanolamine and 28% to polyglycerophosphatides, in *N. europea* 78% belongs to phosphatidylethanolamine and 17% to polyglycerophosphatides (Hagen *et al.,* 1966). In addition, there are bacteria, such as *M. lysodeikticus* and *Staphylococcus aureus,* for instance, which have no phosphatidylethanolamine at all, and the phospholipids consist entirely of mono- and diphosphatidylglycerols (Macfarlane, 1961a, b, and 1962b).

Recently, several bacteria (*Clostridium welchii, S. aureus, B. cereus,* and *Lactobacillus* sp.) have been found to contain o-amino acid esters of phosphatidylglycerol, or lipoamino acids (Macfarlane, 1962b; Houtsmuller and Van Deenen, 1964; Ikawa, 1963). The amino acids in these compounds are mainly lysine, arginine, or ornithine. Lipoamino acids may comprise up to 60% of all the cell phospholipids (Macfarlane, 1964a, b). The maximum lipoamino acid content is found in the stationary phase of development. Lipoamino acids are not found in animal tissues, and many bacteria are devoid of them. Their function in the cell is still obscure.

$$
\begin{array}{c}
\underset{\displaystyle H_2C-O-\overset{\displaystyle O}{\overset{\|}{C}}-R_1}{} \quad CH_2O-\overset{\displaystyle O}{\overset{\|}{C}}-CH-R_3 \\
R_2-\overset{\displaystyle O}{\overset{\|}{C}}-O-CH \qquad \underset{NH_2}{} \\
CHOH \\
H_2C-O-\overset{\displaystyle O}{\overset{\|}{P}}-O-CH_2 \\
OH
\end{array}
$$

Structure of lipoamino acids (Macfarlane, 1962*b*).
R_1, R_2—fatty acid residues; R_3—amino acid residue.

For a long time a difference in lipid composition of gram-positive and gram-negative bacteria was sought as an explanation of the nature of the Gram stain. It was found, however, that no distinct line could be drawn between them. Kates, who reviewed the data on the correlation between the lipid composition of bacteria and their taxonomy, noted that gram-negative bacteria contain more phosphatidylethanolamine and less mono- and diphosphatidylglycerols, whereas in gram-positive bacteria the relative amounts of these groups of phospholipids are the reverse. In addition, several gram-positive bacteria contain lipoamino acids (*o*-esters of phosphatidylglycerols), but gram-negative bacteria rarely contain them (Table 1). The mycobacteria and corynebacteria must be regarded as an isolated group, since their phospholipids consist mainly of phosphatidylinositol glycosides.

Phosphatidylinositol

Table 1. Lipid Composition of Taxonomic Groups of Bacteria (after Kates, 1964)

Family	Major fatty acids	Major phosphatides	Other lipids
Eubacteriales (gram-negative)			
Enterobacteriaceae	16:0; 16:1; 18:1; 17:cy; 19:cy	PE, PS, poly-PG, PG-AA, PG	Glycerides, CoQ
Azotobacteriaceae	16:0; 16:1; 18:1	PE	Glycerides, CoQ
Rhizobiaceae	16:0; 18:1; 17:cy; 19:cy	PE, PC, Me-PE, di-Me-PE	Glycerides, CoQ, poly-β-hydroxybutyrate
Pseudomonadales Pseudomonadaceae	16:0; 16:1; 18:0; 18:1	—	Rhamnolipid
Eubacteriales (gram-positive)			
Lactobacillaceae	16:0; 16:1; 18:1; 19:cy	PG-AA, PG	Glycolipids, glycerides
Bacillaceae (Bacilli)	13:br; 15:br; 17:br	PE, PG, poly-PG, PG-AA	Diglycerides, alcohols, poly-β-hydroxy-butyrate
Clostridia	16:0; 16:1; 18:1; 17:cy; 19:cy	PE, Me-PE, PS, plas-malogen	–
Micrococcaceae	15:br; 17:br	PG, poly-PG, PG-AA	Glycolipids
Corynebacteriaceae	16:0; 16:1; 18:0; higher complex acids	PI-glycosides	Waxes, trehalose esters, glycerides
Actinomycetales			
Mycobacteriaceae	16:0; 16:1; 18:0; 18:1; 19:br; mycolic acids	PI-glycosides	Trehalose esters, mycolic acid esters, peptidolipids

Abbreviations: PE, phosphatidylethanolamine; PS, phosphatidylserine; poly-PG, poly-phosphatidylglycerol; PG-AA, o-amino-acid ester of phosphatidylglycerol (lipoamino acid); PG, phosphatidylglycerol; PC, phosphatidylcholine; Me-PE, phosphatidyl-N-methylethanolamine; di-Me-PE, phosphatidyl-N,N-dimethylethanolamine; PI, phosphatidylinositol. In the column *Major fatty acids*, the first figure is the number of carbon atoms in the chain, the second is the number of double bonds; br signifies a branched chain; cy signifies the cyclopropane ring.

According to recent data, up to 50% of the phospholipids in some mycobacteria consists of diphosphatidylglycerol, 40% of phosphatidylinositol oligomannosides, and 10% of phosphatidylethanolamine (Akamatsu and Nojima, 1965).

Asselineau and Lederer (1960) note the following special features of the neutral lipids of bacteria: absence of sterols, frequent occurrence of branched fatty acids and hydroxy acids, and a high content of free fatty acids. The presence of cholesterol has been reliably established only in mycoplasma (Smith and Rothblat, 1960; Morowitz *et al.*, 1962). From time to time, there have been reports of extraction of traces of sterols from individual species of bacteria (Bultow and Levedahl, 1964), but Lederer quite rightly points out that this amount of sterols does not exceed 0.001% of the dry weight, whereas all other organisms contain 2–7% sterols (Lederer, 1964).

The recent investigations of Erwin and Bloch (1964) showed the absence of di- and polyunsaturated fatty acids. There are only saturated and monounsaturated acids, the amount of which varies in different bacteria. For instance, in *B. megaterium*, monounsaturated acids constitute 4% of all the cell acids, in *M. lysodeikticus* 20%, and in *Corynebacterium diphtheriae* 45% (Fulco *et al.*, 1964). Even bacteria which are very similar morphologically may differ considerably in their fatty acid composition. For instance, out of three cocci (*Gaffnia, Micrococcus criophylus* and *Veilonella alcalescens*), only *Gaffnia* has saturated fatty acid composed of 20 carbon atoms, *M. criophylus* has a monounsaturated acid with 16 carbon atoms, and *V. alcalescens* has one with 17 carbon atoms (Brown and Cosenza, 1964). Some families of bacteria contain cis-11-octadecanoic acid (cis-vaccenic) and related saturated acids containing the cyclopropane ring, mainly cis-11,12-methylene octadecanoic acid (Hoffman, 1962; O'Leary, 1962; Abel *et al.*, 1963).

$$CH_3(CH_2)_5 - \overset{\displaystyle H}{\underset{\displaystyle \diagdown}{C}} - \overset{\displaystyle H}{\underset{\displaystyle \diagup}{C}} - (CH_2)_9COOH$$
$$CH_2$$

cis-11,12-methylene octadecanoic acid (lactobacillic acid)

Lactobacillic acid is found in the phospholipids of *E. coli* (Kaneshiro and Marr, 1961). Mycobacteria are characterized by complex higher fatty acids.

The amount of free fatty acids in bacteria is much greater than in cells of other organisms. In some bacteria (*C. diphtheriae, L. acidophilus, B. megaterium, Agrobacterium tumefaciens, Hemophilus pertussis* and *S. typhimurium*), the amount of free fatty acids constitutes 20% of the total fatty acids (O'Leary, 1962). According to calculations, the free fatty acids in *Sarcina lutea* comprise 2% of the total lipids (Huston and Albro, 1964).

Gram-negative bacteria contain a small amount of glycolipids and relatively large amounts of lipopolysaccharides, whereas gram-positive bacteria are rich in glycosyldiglycerides and contain a small amount of lipopolysaccharides. As an example we can mention mannosylmannosyl (1 → α)-diglyceride, isolated from cells of *M. lysodeikticus* (Lennarz, 1964).

Table 2.   Composition of Bacterial Membranes (% Dry Weight)

Bacteria	Proteins	Lipids	Nucleic acids	Carbohy- drates	Author
Micrococcus lysodeikticus	50	28	0.2	15–20 (mannan, hexosamine)	Gilby, Few, and McQuillen, 1958
Sarcina lutea	40	29	3.6	10	Brown, J., 1961
Bacillus megaterium M.	63–69	16–20	1.5	1–10 (as glucose)	Weibull and Berg- strom, 1958
Bacillus megaterium M.	—	—	11	—	Vennes and Ger- hardt, 1956
Bacillus megaterium M.	50	26	24	2	Godson *et al.,* 1961
B. megaterium KM	75	23	—	·1	Yudkin, 1962
Staphylococcus aureus	41	22.5	—	2	Mitchell and Moyle, 1956, 1959
Pseudomonas fluorescens	50	16–18	5–7	—	Hunt *et al.,* 1959
Mycobacterium laidlawii	—	25–30	—	7–8	Smith, P., 1964
Streptococcus A	—	25	—	—	Freimer and Krause, 1960
Streptococcus faecalis ATCC	—	21	—	—	Ibbott and Abrams, 1964a, b
Streptococcus faecalis	49–55	28	4	0.1	Shockman *et al.,* 1963
Mycoplasma sp.	—	36	—	—	Razin *et al.,* 1963

Although a knowledge of the lipid composition of intact bacterial cells may make some contribution to an understanding of the composition of membranes, information obtained from analysis of isolated membranes is of greater value. The lipid content of membranes of various bacteria reaches 30% (Table 2), which is the same as that of mitochondria (Ball and Joel, 1962). There is a relationship between the lipid content of membranes and the stage of development of the culture. For instance, the membranes of *S. faecalis* contain 28% lipids in the exponential phase and up to 40% in the stationary phase (Shockman *et al.,* 1963).

The large amount of carbohydrate components in the membranes of *M. lysodeikticus, S. lutea, B. megaterium,* and *M. laidlawii* is probably due to the presence of polysaccharides of the mannan type. The ratio of phospholipids to neutral lipids is not constant in bacterial membranes. For instance, in *B. megaterium* KM, more than 50% of the membrane lipids consist of neutral lipids (Yudkin, 1962), while in *M. lysodeikticus* neutral lipids comprise about 20% (Macfarlane, 1961*a, b*). Mitochondria, on the contrary, contain these components in a more constant ratio (about 90% phospholipids and 10% neutral lipids). The phospholipids of the membranes of grampositive bacteria have been most thoroughly investigated (Table 3). The results of these investigations agree with the data obtained from an analysis of the phospholipids of intact cells.

It has recently been established that the phospholipids of the membranes of some gram-positive bacteria contain both mono- and

Table 3.   Composition of Membrane Phospholipids

	Bacillus megaterium		*Micrococcus lysodeikticus‡*	*Streptococcus faecalis§*
	M*	KM†		
Mono- and diphosphatidylglycerols	90	—	76	70
Phosphatidylethanolamine	0	97	0	—
Phosphatidylinositol	0	—	12.9	—
Glycolipid	—	—	About 10	30

* Weibull, 1957.
† Yudkin, 1962, 1966; the results for these two strains are expressed as % of the weight of the total phospholipids (Kodicek, 1963).
‡ Macfarlane, 1961 (P content as % of total P of phospholipids).
§ Ibbott and Abrams, 1964*a,b* (% by weight of total phospholipids).

diphosphatidylglycerols. In *S. faecalis* phosphatidylglycerol domi-
nates, with a little phosphatidic acid and no diphosphatidylglycerol at
all (Vorbeck and Marinetti, 1965). This agrees with the results of an-
alyses by Shockman *et al.* (1963), who earlier suggested a similarity
between the composition of the membranes of *S. faecalis* and those of
M. lysodeikticus and *B. megaterium* M as regards predominance of
phosphatidylglycerols. But there are other data indicating a predominance of diphosphatidylglycerol. These differences in the phospholipid composition may be due to differences in the age of the culture. The
phospholipids of the membranes of *M. lysodeikticus* have been
thoroughly investigated. Macfarlane found that phosphatidylglycerol
and diphosphatidylglycerol made up 76% of the total lipid phosphorus. Out of this, 72% consisted of the phosphorus of phosphatidylglycerol and 4% of diphosphatidylglycerol (Macfarlane, 1961*b*), but
this ratio could change in favor of the latter with appropriate cultivation conditions (Macfarlane, 1961*a*). The fatty acids in these phospholipids consisted mainly of: (1) 12-methyltetradecanoic; (2) phosphatidylinositol, 12.9% of total phospholipid phosphorus; (3) a
glycolipid containing a fatty acid ester, carbohydrate, and phosphorus in molar ratio 5.7:3.2:1.0. No lipoamino acids could be found
in the cells or membranes of *M. lysodeikticus* (Macfarlane, 1964*a*).

According to Ibbott and Abrams (1964*b*), a glycolipid containing phosphorus is present in the membranes of *S. faecalis*. The ester:
phosphorus ratio is 6:1.

As already mentioned, *o*-amino acid esters of phosphatidylglycerol are found in intact cells of several bacteria. Vorbeck and
Marinetti (1964*a, b*) recently established that the *o*-ester of lysine
and phosphatidylglycerol was present in the membranes of *S.
faecalis*.

These compounds have not yet been isolated from the membranes of other bacteria but one may be sure it is in the membranes,
where phospholipids, particularly phosphatidylglycerol, are always
concentrated that they will be found. Hunter and Goodsall (1961)
give evidence of the presence of lipoamino acids in the membranes
of *B. megaterium* KM. It is very probable that the high nitrogen:
phosphorus ratio found in the membrane lipids of *S. aureus* (Few,
1955) and *B. megaterium* KM (Yudkin, 1962) can be attributed not
only to the postulated presence of phosphatidylethanolamine and
phosphatidylserine, but also to the presence of lipoamino acids. A
final conclusion will have to be based on quantitative determinations
of the lipoamino acid content of the membranes. This is difficult

since they are easily hydrolyzed by the membrane enzymes while the preparations are being obtained.

Despite the fact that the phospholipid composition has not been well investigated even in the membranes of gram-positive bacteria, the available data indicate great qualitative differences not only between the large taxonomic groups, but even between related species. As an example, we can mention strains M and KM of *B. megaterium*. In the first strain, the membranes contain no phosphatidylethanolamine nor lipoamino acids, and the phospholipids consist entirely of mono- and diphosphatidylglycerols, whereas in the second phosphatidylethanolamine predominates, lipoamino acids are present, and the amount of acid phosphatides is very small (Table 3). Changes have been observed in the phospholipid composition of membranes of *B. megaterium* strain MK 10D in relation to the composition of the nutrient medium (Op den Kamp *et al.*, 1965). Similar observations have been made for *S. aureus* cells, where an increase in acidity of the medium increased the content of the lysine ester of phosphatidylglycerol (Houtsmuller and Van Deenen, 1965).

Akamatsu *et al.* (1966) managed to obtain a fraction composed of fragments of the internal membranes of *Mycobacterium phlei*. Diphosphatidylglycerol accounted for 50%, phosphatidylinositol oligomannosides for 37%, and phosphatidylethanolamine for 10% of the membrane phospholipids.

Thus, the special features of the phospholipid composition of bacterial membranes include the rare occurrence of phosphatidylcholine, a variable ratio of phosphatidylethanolamine to phosphatidyglycerols, the presence of lipoamino acids and glycolipids, and also the variation of the quantitative and qualitative composition with the conditions of growth and age of the culture. In contrast to this, the phospholipid composition of mitochondria from various different cells is generally similar and consists of a set of several phospholipids. For instance, beef-heart mitochondria contain phosphatidylcholine (37%), phosphatidylethanolamine (31%), phosphatidylinositol (10%), and diphosphatidylglycerol (16%) (Fleischer *et al.*, 1961). However, the individual phospholipids of mitochondria differ from one another in fatty acid composition (Macfarlane, 1964*a*). For instance, a typical phosphatidylcholine contains 50% saturated acids and a small amount of polyenoic acids. In the phosphatidylinositol from heart or liver, stearic acid comprises 50%, there is a little linoleic acid, and polyenoic acids constitute up to 30%. The diphosphatidylglycerol contains up to 80% unsaturated acids, mainly linoleic, and about 20% oleic acid. In contrast to this, the phospholipids of *M. lysodeikticus* consist mainly of C_{15}-branched acids with a high content of 12-methyltetradecanoic (anteiso) acid. In *S. aureus* the phospholipids contain up to 70% branched acids (mainly C_{15}) and about 20% normal saturated acids (C_{16}, C_{18}, C_{20}) (Macfarlane, 1964*a*). The phospholipids of *B. megaterium* contain the saturated C_{15} fatty acids 12-methyltetradecanoic and 13-methyltetradecanoic (Yudkin, 1966).

In addition, the diphosphatidylglycerol from *S. faecalis* contains five different fatty acids: 14:0, 16:0, 16:1, 18:1 (cis) and 19:0 cyclopropane (Ibbott and Abrams, 1964*b*).

Unfortunately, there are no reliable data on the phospholipid composition of the membranous systems of gram-negative bacteria. This is because the membranous structures cannot be separated completely from the lipoprotein membrane of the cell wall. The importance of phospholipids in membrane formation in gram-negative bacteria has been shown in the example of chromatophore formation in *R. spheroides* (Lascelles and Szilagyi, 1965). The membranous apparatus of this photosynthetic bacterium contains 157mμmole of lipid phosphorus per milligram of protein, while the bacterium grown aerobically in darkness contains 90 mμmole. Enrichment of the membrane with phospholipids increases the solubility of chlorophyll.

Kates (1964) advanced the interesting suggestion that the membrane phospholipids of all bacteria are alike and consist mainly of phosphatidylglycerols. The excess amount of phosphatidylethanolamine in most gram-negative bacteria is due to the membrane of the cell wall. In this connection the data indicating a role of phosphatidylethanolamine in the synthesis of cell-wall lipopolysaccharides in *S. typhimurium* (Rothfield and Horecker, 1964) are of interest. Kates' hypothesis is based on the essential presence of phosphatidylglycerols in biologically active membranes involved in energy-transformation processes—mitochondria and chloroplasts. The membranes of the endoplasmic reticulum contain hardly any phosphatidylglycerol (Macfarlane, 1964*a*).

The neutral lipid fractions of mitochondria contain cholesterol, ubiquinone, and a small amount of carotenoids and glycerides (Green and Fleischer, 1963). The presence of free fatty acids and lysophosphatides in some mitochondrial preparations is believed to be due to lipid degradation on disintegration of the cell (Ball and Joel, 1962).

What distinguishes the neutral lipid fraction of bacterial membranes from that of mitochondria?

Bacterial membranes contain no cholesterol. Exceptions are some mycoplasma species, the cytoplasmic membrane of which contains up to 70% of the total cell cholesterol (Tourtellotte *et al.,* 1963; Argaman and Razin, 1965). According to Van Deenen (1964) and Van Deenen and Van Golde (1966), cholesterol stabilizes the various biological membranes by the formation of esters with the phospholipids. This interaction of phospholipids with cholesterol may occur in the membrane of mycoplasma, which lacks a cell wall (Smith, P., 1964). In this connection it is of interest that some strains of *Mycoplasma* form a cytoplasmic membrane only when cholesterol is present in the nutrient medium (Rodwell and Abbott, 1961). According to the data of some authors, the cell contains two forms of cholesterol. One is a dynamic form, which is easily esterified and takes part in the transport of fatty acids. The second is a structural form, which is involved in the structure of the cell membranes (Bourne, 1958). There is evidence that in bacteria which lack cholesterol, some of cholesterol's functions may be performed by carotenoids (Smith, P., 1963). In Smith's opinion, carotenoids in bacteria, like sterols in mitochondria, are responsible for the structure of the membranes and bring about transport of fatty acids and sugars through the membrane by the formation of glycosides or carotenoid esters. According to Razin *et al.,* (1966), however, the resistance of *Mycoplasma laidlawii* to osmotic shock does not depend on cholesterol or carotenoids, but on the unsaturated fatty acids in the phospholipid molecules.

The presence of carotenoids and hydroxycarotenoids in bacterial membranes has been demonstrated in several investigations (Mathews and Sistrom, 1959; Stephens and Starr, 1963; Jackson and Lawton, 1958; Sasaki, 1964).

P. Smith (1963) found carotenoids of the neurosporene (tetrahydrolycopene) type and hydroxycarotenoids in the free state and in the form of a fatty acid ester in *Mycoplasma* membranes. Hydroxycarotenoids have also been found in the membranes of other bacteria. For instance, Gilby and Few (1958) found a carotenoid in the membranes of *M. lysodeikticus* and postulated that it had a structure of the lutein type. Further analysis established the presence of seven carotenoids in *M. lysodeikticus.* These had the same chromophore group but differed in polarity. It has been suggested that they are hydroxylated carotenoids of the neurosporin type (Rothblat *et al.,* 1964).

Bacterial membranes contain various quinones—derivatives of vitamin K (naphthoquinones) and coenzyme Q (ubiquinones) with different side-chain length. The quinone content reaches 0.5–1 μ mole per gram of dry weight. The quinones of bacterial membranes will be discussed more fully in connection with the enzymatic composition of the respiratory chain (Chapter III).

The membrane glycerides have received little study. It is known that a diglyceride is the main component of the neutral lipids of the membranes of *M. lysodeikticus* and 77% of this diglyceride is a C_{15} acid (Macfarlane, 1961a). A glucosyldiglyceride and a glucosyl-galactosyldiglyceride have been found in the membranes of *S. faecalis* (Vorbeck and Marinetti, 1965).

Fatty acids form part of the molecules of phospholipids and neutral lipids, and the structure of these hydrocarbon chains determines to some extent the structure and strength of the biological membrane. In addition, the membrane may also contain free fatty acids. It is of interest to note several features of the fatty acid composition of bacterial membranes. As Table 4 shows, bacterial membranes usually lack di- and polyunsaturated fatty acids, and the amount of monounsaturated acids is insignificant, with some rare exceptions. Branched fatty acids and acids with the cyclopropane group in their chain are typical of the lipids of bacterial membranes. In the membranes of *L. casei,* lactobacillic acid comprises about 17% of the total fatty acids. C_{17} and C_{19} fatty acids containing cyclopropane in their chain have recently been found in membranes from *S. faecalis* (Ibbott and Abrams, 1964a, b). The fatty acid composition of mitochondrial lipids differs significantly from that of bacterial-membrane lipids. Mitochondria contain no acids with the propane ring, a small amount of branched acids, and a far greater amount of unsaturated acids than bacteria.

In discussing the lipid composition of bacterial membranes, we have repeatedly stressed the differences between different species of bacteria and between bacteria in general and higher organisms. It is natural to consider how the differences in the lipid composition of bacterial membranes and mitochondria are linked with their structure. In bacterial structures the major element is a unit membrane 75 Å in diameter. In the course of evolution, its "packing" has been altered, and double membranes and complex structures (mitochon-

Table 4. Fatty Acid Composition of Bacterial Membranes and Cell Walls (as % of Total Fatty Acids)

Acid	Lactobacillus casei	Micrococcus lysodeikticus	Bacillus megaterium	Streptococcus faecalis	B. licheniformis	Bacillus stearothermophilus	Sarcina lutea	Aerobacter aerogenes	Escherichia coli	Pseudomonas sp.	S. gallinarum
Less than 12	0.1	2.9	2.8	—	—	—	—	—	—	—	—
12:0 unbr	0.3	0.5	0.7	0.5	—	—	—	—	—	2.6	2.9
12:1	0.2	0.8	1.1	—	—	—	—	—	—	—	—
13:0 unbr	0.2	0.3	0.5	0.1	—	—	—	—	—	—	5.9
14:0 br	0.5	1.1	3.0	—	4.2	2.9	6.0	—	6.8	3.5	9.1
14:0 unbr	2.2	0.8	2.6	2.2	—	4.2	1.8	5.7	2.1	2.9	3.8
14:1	0.8	0	0	—	—	—	—	—	—	—	—
15:0 br (iso)	0.4	71.7	26.0	—	—	—	—	—	—	—	—
15:0 br (anteiso)	0.5		29	—	50.4	30.7	80.2	—	—	—	—
15:0 unbr	0.9	1.5	2.9	0.6	—	—	—	—	0.8	4.2	3.8
16:0 br (iso)	0.5	0.6	1.3	0.4	—	—	—	—	—	—	—
16:0 br (anteiso)	0	1.7	2.1	—	—	—	—	—	—	—	—
16:0 unbr	21.5	5.1	7.3	29.6	3.9	3.1	1.5	56.4	42	49.5	49
16:1	7.2	2.0	3.4	16.4	12.0	23.5	1.3	5.7	2.3	18.3	4.7
17:0 br	1.1	3.3	2.6	1.5	28.2	33.2	1.3	—	—	—	—
17:0 unbr	1.6	0.2	0.4	—	—	—	—	—	—	—	—
17:1	4.5	0.6	1.3	—	—	—	—	3.8	—	8.1	—
18:br (iso)	1.1	0.6	0.7	—	—	—	—	—	—	—	—
18:0 unbr	6.8	2.3	2.6	11.8	0.2	1.2	2.5	—	1.7	8.6	2.0
18:1	21.7	2.4	8.8	21.0	1.1	1.2	3.6	9.6	25.7	2.3	2.0
18:2	—	—	—	—	—	—	—	18.8	—	—	—
19:br (anteiso)	3.0	0	0	—	—	—	—	—	—	—	—
19:propane	16.0	0	0	—	—	—	—	—	—	—	—
20:1	0	1.2	0	—	—	—	—	—	—	—	—
21:0 unbr	8.3	0.4	2.0	—	—	—	—	—	—	—	—

NOTE. *Lactobacillus casei, Micrococcus lysodeikticus, Bacillus megaterium*, according to Thorne and Kodicek, 1962; *Streptococcus faecalis*, according to Freimer, 1963; *Bacillus licheniformis, Bacillus stearothermophilus, Sarcina lutea, Aerobacter aerogenes, Escherichia coli, Pseudomonas sp.*, and *Salmonella gallinarum*, according to Cho and Salton, 1964 and 1966.

dria) separated from the cytoplasm by a special membrane have appeared. It would seem that the lipid components which we find in mitochondria are not all essential for the construction of a unit lipoprotein membrane. The earliest unit membranes could probably have arisen when a particular protein:lipid ratio was established, provided that the hydrophilic and hydrophobic residues, as well as the acid and basic groups in phospholipids and proteins, were in the required proportions, and that the branches or rings in the fatty acid residues had the necessary spatial arrangement. Membranes of such type promoted the separation of the first living cells and have been retained in some form or another in bacteria.

The comparatively simple lipid composition and uniform fatty acid composition of the lipids in each particular kind of cell have been replaced in the process of evolution by a variety of lipids, with a predominance of phosphatidylcholine, cholesterol, and unsaturated fatty acids. These radical changes in lipid composition and, in particular, in phospholipid composition, associated, of course, with the development of new biosynthetic mechanisms, probably gave rise to more complex membranous systems with a more stable structure, and subsequently to mitochondria.

To give a clearer illustration of the spatial organization of bacterial membranes we will consider briefly their ultrastructure and the role of protein–lipid interactions.

BASIC PRINCIPLES OF BACTERIAL MEMBRANE CONSTRUCTION

The most general properties of lipids, which account for their role in biological structures, are the ability to arrange themselves in an ordered manner, particularly to form bimolecular layers (lamellae) or micelles, and the ability to interact with proteins to form lipoproteins. Lipoprotein complexes are probably the main form of spatial organization for enzyme systems, where the necessary conformation for activity of the enzymes and the interaction of the enzymes with one another are effectively secured.

To understand the ultrastructure of bacterial membranes, it is important to know these properties of lipids; hence, we discuss the question briefly, with emphasis on the phospholipids which predominate in biological membranes. Until recently it was believed that phospholipids in biological and model membranes are arranged in the form of a bimolecular lamellar layer (Davson and Danielli, 1943; Robertson, 1959, 1961; Stoeckenius, 1959). Phospholipid molecules have polar groups (the fourth nitrogen atom, the phosphate ion) and nonpolar hydrocarbon chains of fatty acids. In a bimolecular layer of lipid, the hydrocarbon chains are inside, while the polar groups point towards the surface, where they may be connected with water dipoles, polar groups of proteins or polysaccharides (Fig. 8a and d). Stoeckenius (1964) believes that the fatty acid residues which constitute the

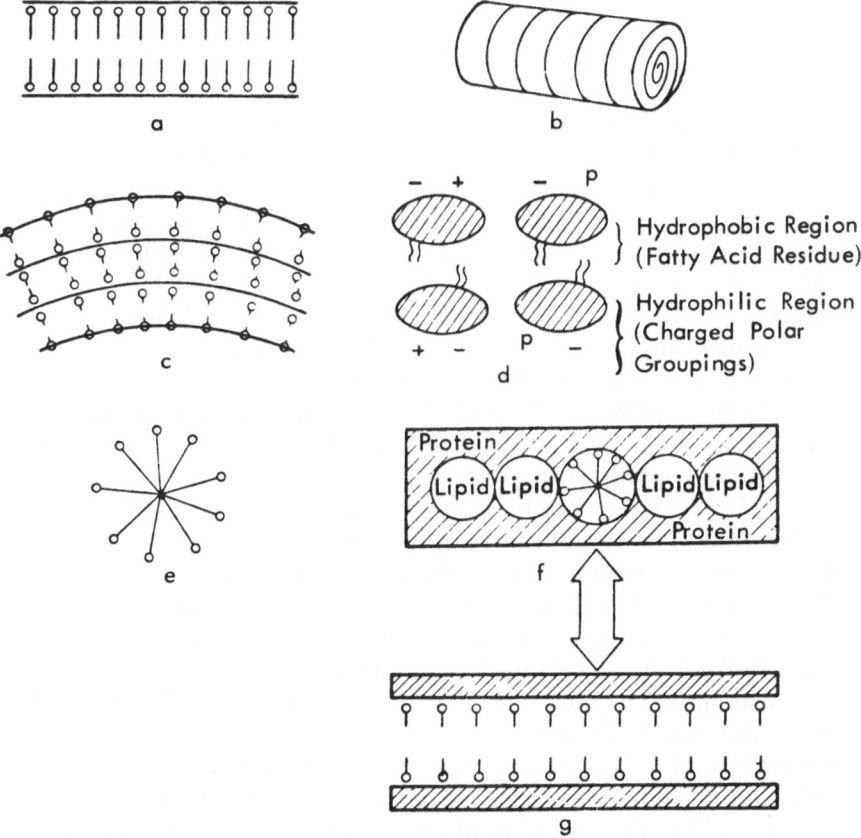

Fig. 8. Types of lipid membrane structure: (a) phospholipid lamella; (b) lecithin micelle; (c) lecithin micelle constructed of lamellae; (d) segment of lamella; (e) phospholipid micelle; (f) structure of membrane with lipid in micellar form; (g) structure of membrane with lipid in lamellar form (b,c,d from Green and Fleischer, 1962; e,f,g from Lucy, 1964).

internal part of the lipid lamella are in an almost liquid state and freely rotate and bend.

Some authors have shown that, in addition to the lamellar arrangement of lipid in a bimolecular layer, there is a micellar form (Bangham and Horne, 1964; Lucy and Glauert, 1964; Lucy, 1964; Kavanau, 1963). Small lipid micelles (Fig. 8e) can aggregate and form a continuous system of lipid micelles with their polar surfaces in contact and, as in lamellae, may be covered with protein (Fig. 8f). The stability of the micellar form of phospholipids depends on the nature of the charge, the degree of unsaturation of the fatty acids, and other factors (Fleischer and Klouwen, 1961). The lamellar and micellar forms of organization of phospholipids are very labile and each form can change to the other. Model experiments have shown that the lamellar structure may be destroyed if electrostatic, surface, or intramolecular interactions are altered. For instance, in Bangham and Horne's experiments the introduction of an unbalanced charged group (an anion or cation with a long chain) led to the partial or complete micellization of lamellae. Lamellae of phosphatidylcholine and cholesterol are converted into micelles 70–80 Å in diameter by the introduction of lysophosphatidylcholine or saponin. This can be observed with the aid of negative-contrast staining. The lamella-micelle transition can be brought about by the interaction of detergents with the membrane (Seufert, 1965).

Many properties of biological membranes can be explained on the assumption of the occurrence of a lamella–micelle transition when both types of organization are present in the same membrane (Fig. 8f and g) (Lucy, 1964). The ability of phospholipid micelles to alter their configuration when conditions in the ambient medium change is the main property which determines their fundamental role in the formation of biological membranes. In this connection Lehninger's (1964) idea of the existence of two or three thermodynamically stable conformations of the membrane is of interest.

Thompson (1964a) thinks that the structure of bimolecular lipid layers (lamellae) depends on the interactions which are responsible for the secondary, tertiary, and quaternary structure of protein molecules. The stability of these layers depends on the nature and relative amounts of the lipid components, their electrical charges, and steric properties.

In view of the increasing interest in the role of phospholipids in biological membranes, model experiments have been conducted on lipid–water systems to study the spatial arrangement of lipids. These experiments have shown that, although in principle the orientation of lipid molecules is constant within a certain range of lipid:water ratios, structures of various degrees of complexity, in addition to simple lamellae composed of two monomolecular layers of lipid, can be formed, depending on the nature and concentration of the lipid, temperature, and other factors. For instance, solubilized lecithin forms a complex lamellar system (Fig. 8b and c) in which the individual bimolecular layers are densely packed and folded in a particular manner (Green and Fleischer, 1962, 1963).

Of course, information on the structure of model phospholipids and mixed lipid membranes can be extended to biological lipoprotein membranes only after appropriate investigation. Yet the ideas of the lamellar and micellar structure of phospholipids, their interconvertibility, and the inhomogeneous structure of the biological membrane are very attractive, since they explain several functions of the membrane (permeability, organization of respiratory-chain components) better than the idea of a uniform lipid layer. The idea of a static and constant biological membrane will probably be replaced by the concept of highly dynamic membranes in which the ultrastructure is continually changing (Kavanau, 1965; Hechter, 1965; Van Deenen, 1965).

Green notes that phospholipids can be obtained in micellar form only if their acids are unsaturated. He refers to the relatively high amount of unsaturated fatty acids in mitochondrial membranes and suggests that the phospholipids in them are also arranged in the form of oriented micelles (Green and Fleischer, 1962, 1963). In fact, 68% of mitochondrial fatty acids are unsaturated and 40% of them have two or more double bonds in the molecule (Ball and Joel, 1962). It has been known for a long time that polyunsaturated fatty acids are important constituents of the animal diet. They are found in all membranous organelles, in erythrocytes, and in the endoplasmic reticulum. The absence of polyunsaturated fatty acids leads to defective permeability. It has been shown in model experiments that the degree of unsaturation of a fatty acid also affects the size of the micelle, its shape and, in particular, its ability to combine water-insoluble molecules. The mechanism of this effect is unknown. The fact that many water-insoluble compounds such as coenzymes Q and vitamin K are solubilized by phospholipid micelles suggests that it is this property of phospholipid micelles that is responsible for one of the biological functions of lipids—the transport of water-insoluble material (Green and Fleischer, 1962). Oxidation of unsaturated fatty acids hinders the formation of model membranes from various phospholipids (Huang *et al.,* 1964). One of the difficulties encountered in the attempt to extend the theory of lamellar–micellar structure to bacterial membranes is the absence of di- and polyunsaturated acids in bacterial phospholipids. However, according to Kodicek and Kates, unsaturated fatty acids are not essential for the construction of biological membranes. Membranes can be constructed from branched acids and acids containing the propane ring in their chain. Such fatty acids are common in bacterial membranes and are present in considerable amounts. They also impart flexibility and elasticity to the bacterial membrane when the components are densely packed (Kodicek, 1963; Kates, 1964).

An important difference between bacterial and mitochondrial membranes is the absence of phosphatidylcholine and cholesterol in bacteria. These components are considered to be largely responsible for the structure of mitochondrial membranes, which changes relatively easily from the lamellar form to the micellar form. In a number of bacteria the role of phosphatidylcholine is probably played by phosphatidylethanolamine. Many bacteria, however, have no amphoteric phospholipids at all and contain only phosphatidylglycerol and diphosphatidylglycerol. In this case the organization of the membrane lipids is difficult to illustrate in model experiments, since it has been impossible so far to obtain lamellae or micelles of pure diphosphatidylglycerol (Huang *et al.,* 1964).

A significant drawback of investigations using model lipid membranes to study the role of phospholipids in their lamellar or micellar form is that the effect of interactions with protein on the ultrastructure and stability of such model structures has not been taken into account. In biological membranes, lipids occur only in a complex with proteins. Hence, it is of interest to determine the interactions involved in the construction of lipoproteins.

Most lipids in the bacterial cell can be extracted with neutral solvents, but 5–20% cannot be isolated except after hydrolysis. This indicates a covalent link of lipids with proteins. Yet the structure of most lipoproteins is not based on covalent forces, but on weaker forces—long-range or adsorption forces (Salem, 1962, 1964; Ball and Joel, 1962; Cook and Martin, 1962; Fleischer *et al.,* 1962; Bresler, 1963).

Lipids containing polar groups can be bound with protein by electrostatic forces. Phospholipids interact with protein via their phosphate groups and/or the fourth nitrogen atoms (lecithins and cephalins); free fatty acids react via their carboxyl groups. These lipids show an affinity for amino acids containing the groups —OH, $>$NH, —NH$_2$, and $>$S. Polarization forces due to polarization of one molecule by the charges of another play an important role. This gives rise to an interaction between the polar group of one molecule and the polarized group of another. For instance, the —CH$_2$ group is polarized by the charge of the phosphate ion at a short distance, and this gives rise to a dipole, which is attracted to the phosphate ion. London–Van der Waals forces are also of great importance. These are quantum-mechanical dispersion forces, which arise between methylene groups of adjacent hydrocarbon chains. Kauzmann (1959) thinks that dispersion forces tend to bring together identical groups, whether they be aromatic or aliphatic groups. The energy of one bond of this type is 0.1 kcal/mole. Even such a small energy may lead to strong attractions owing to the large number of similar parallel interactions.

Van der Waals forces are of special significance in the interaction of nonpolar groups, since in this case there is no possibility of any other forces acting (hydrophobic interactions, according to Green). Biochemists have introduced the term "hydrophobic bond" to denote the tendency of nonpolar regions of the molecules of proteins, nucleic acids, or lipids to avoid water and combine (Sinanoglu and Abdulnur, 1965).

Thus, the hydrophobic bond is the result of strong dispersion attraction between nonpolar groups and the strong electrostatic attraction between water molecules or between them and polar groups. Generally speaking, it is believed that water molecules packed in a particular manner play an important role in the construction of biological lipoprotein membranes (Klotz, 1965; Ning Ling, 1965; Hechter, 1965; Salem, 1964).

In the opinion of Fleischer et al. (1962), the hydrogen bond may be formed in the interaction of phospholipid micelles with proteins, but the relative contribution of this bond, like that of the covalent bond, is small.

Since lipid–protein interactions are fundamental for maintenance of the structure of biological membranes, it would be of great interest to know the relative importance of each kind of bond in the construction of various biological membranes. There is some information on this point for mitochondrial membranes. There are model experiments which indicate electrostatic (ionic) interactions in the construction of mitochondrial membranes. One example is the bond between the basic protein, cytochrome c, and phospholipids which have an excess of negative charges (phosphatidylglycerol and phosphatidylinositol). Another example is the interaction of nitrogenous phospholipids (phosphatidylcholine and phosphatidylethanolamine) with proteins having an excess of negative charges (Green and Fleischer, 1963).

Hydrophobic (nonionic) interactions play an important role in the stabilization of mitochondrial membranes. The fatty acid residues of the phospholipids form complexes with the amino acid residue of enzymatic and structural proteins in which hydrophobic side chains predominate (isoleucine, leucine, methionine and valine).

In a comparative analysis of group C cytochromes from vertebrates and yeasts, Smith and Margoliasch noted a predominance of amino acid residues with hydrophobic side chains, which could reach 41% (tryptophan, tyrosine and phenylalanine). In their opinion, these regions of the molecule serve for hydrophobic interactions with other components of the respiratory chain. On the other hand, the large amount of the basic amino acid (lysine) imparts a basic character to the whole cytochrome c molecule and allows it to enter into ionic interactions (Smith and Margoliasch, 1964; Smith and Minnaert, 1965). These important properties of the cytochrome c molecule have been elaborated, in these authors' opinion, in the process of evolution.

We know much less about the bonds that predominate in the structure of the lipoprotein complexes of bacterial membranes. An investigation of purified bacterial cytochromes of group C failed to show their homology with animal proteins. Many bacterial cytochromes of group C are acid proteins, as has been shown for cytochromes c_4 and c_5 from *Azotobacter vinelandii,* for cytochromes c from two species of *Pseudomonas,* and for several other bacterial cytochromes (see Table 11) (Tissières, 1956; Tissières and Burris, 1956; Horio, 1958a, b; Ambler, 1963; Coval *et al.,* 1961). Hence, in model experiments, they were incapable of complexing with phospholipids with an excess of negative charges (Green and Fleischer, 1963). The apparent predominance of acid proteins (cytochromes) in bacterial membranes, together with the fact that basic enzymes or structural protein have not yet been found there, raises a question as to the nature of the interactions between proteins and lipids which form a stable membrane in bacteria. Although it is fairly easy to conceive the nature of electrostatic interactions with proteins in a number of gram-negative bacteria where amphoteric phospholipids (phosphatidylethanolamine) predominate, in many gram-positive bacteria, where acid phospholipids (phosphatidylglycerol, phosphatidylinositol) predominate, it is difficult to postulate the formation of a strong membrane by interaction with acid proteins.

The bacterial membrane, which contains many acid phospholipids and is poor in basic proteins, obviously requires stabilization to balance the excess negative charges. It is possible that o-amino acid esters of phosphatidylglycerol, which are found mainly in gram-positive bacteria, play some role in compensation of the excess negative charges. These compounds, however, are not present in all bacteria or at all stages of development and only appear when the acidity of the medium is raised (Macfarlane, 1964a; Houtsmuller and Van Deenen, 1964; Van Deenen, 1964).

Stabilization of the bacterial membranes involves bivalent cations, including magnesium and calcium, which impart strength to the bacterial membrane. The cations are probably the bridges which neutralize the excess of negative charges and bind acid phospholipids and acid proteins together (phospholipid–R–COO–Mg–OOC–R–protein). This idea is supported by facts which indicate an effect of EDTA (Versene) on membranes obtained from *M. lysodeikticus*. Treatment with $10^{-3}-5 \times 10^{-3}\ M$ Versene leads to destruction of the membranes (Lukoyanova *et al.,* 1961). Data on the role of cations have been obtained for other bacteria, including marine bacteria (Brown, A. *et al.,* 1962; Brown, A., 1962a, b; Walter and Eagon, 1964; MacLeod, 1965; Brown, A. *et al.,* 1965). In Brown's opinion, the removal of cations results in conformational changes in the membrane proteins and the appearance of protease activity. This process probably leads to the loss of the binding cationic bridges and to an increase in hydrophilia due to the liberated carboxyl groups. This results in destruction of the membranes. It has been found that carboxyl groups are located on the surface of the membrane of halophilic bacteria, while the basic groups lie inside the membrane (Brown, A., 1956). In this connection, an increase in the carboxyl groups in the membranes of *Pseudomonas* NC MB 845 by succinylation was accompanied by an increase in hydrophilia, and more concentrated salt solutions were required for stabilization of the membrane (Brown, A., 1964*b*). A reduction of hydrophobicity and a concomitant increase in the solubility of proteins due to succinylation were also found in an investigation of fragments of mitochondrial membranes (MacLennan *et al.*, 1965). J. Brown (1965) attempted to investigate the mechanism of membrane stabilization by Mg^{++}. He found that sonication of membranes in a solution containing Mg^{++} gave 70S and 5S particles. On subsequent removal of Mg^{++} by dialysis, the 70S particles broke up into 5S particles. The membrane apparently consists of relatively small liproprotein aggregates (5S particles), bound together by Mg^{++}. Lysis of some bacteria (*P. aeruginosa* 64, *A. faecalis*) can be effected by ethylenediaminetetraacetic acid in the absence of lysozyme (Eagon and Carson, 1965; Gray and Wilkinson, 1965).

According to some authors, Versene in a concentration of 10^{-3} *M* does not act on the mitochondrial membrane, i.e., the relative importance of this kind of interaction with cations in these membranes is not great. In view of their composition, proteins and phospholipids can interact directly with one another and cationic bridges are not necessary. Versene is sometimes used even as a stabilizer of mitochondrial membranes (Slater and Cleland, 1952; Skulachev, 1962). In addition, it has recently been shown that in some cases 5×10^{-5} *M* Versene has no effect on the state of mitochondrial membranes or the operation of the respiratory chain (Arcos and Argus, 1964).

Polyamines (spermine and spermidine) have a stabilizing effect on bacterial membranes (Tabor *et al.*, 1961; Tabor, 1962; Grassowicz and Ariel, 1963). The mechanism of this effect has not been worked out, but there are grounds for assuming that the polyamine interacts with the acid groups of the membrane and reduces its hydrophilia (Brown, A., 1964*a*; Harold, 1964). The interaction of a polyamine with the membrane changes its architectonics in a particular manner and may lead to death of the cell. There are reports that the interaction of protamine with membranes of intact *E. coli* or *S. aureus* cells leads to changes in permeability to glucose and phosphate (Kuznetsov, 1965). A stabilizing effect of bivalent cations (Mg^{++}, Fe^{++}, Ba^{++}, Ca^{++}) and polyvalent cations (spermine, spermidine) has been observed in the case of the cell membrane of *Mycoplasma*.

The lowest concentration of magnesium ions which has a stabilizing effect in *Mycoplasma* is 10^{-5} *M.*

There is very little direct evidence of the occurrence of hydrophobic interactions in bacterial membranes. In a recent paper, A. Brown (1965) showed that the lipoprotein complexes of the membrane of halophilic bacteria are maintained by hydrophobic and hydrogen bonds. The organization of these complexes into a membrane is effected mainly by electrostatic bonds.

From data describing the effect of detergents on gram-positive and gram-negative bacteria and mitochondria, we can surmise that the relative importance of hydrophobic interactions is fairly great in all cases. Bacterial membranes, like mitochondrial membranes, are destroyed by detergents.

Detergents consist of molecules which have a hydrophobic hydrocarbon radical and a hydrophilic, solubilizing or polar group with strong valence forces (Schwartz and Perry, 1953). Anionic, cationic, and nonionic detergents can be distinguished. The classes of surface-active substances (typical structures) are given on p. 58, according to Newton (1960).

The mechanism of action of these substances consists in the formation of a monolayer at the phase boundary owing to the ability of the molecules to adopt an oriented position; this reduces the surface tension. One must distinguish between the action of detergents on isolated proteins and on biological structures. Their action in the latter case is thought to be due to rupture of hydrophobic connections, whether they are protein–protein or lipid–protein bonds. As an illustration of the change in mitochondrial structure due to the action of detergents we can refer to the loss of cristae due to the action of 0.1 *M* caprylate (Adams *et al.,* 1963).

Investigations of the effect of detergents on the membranous structures of bacteria show that the detergent destroys the lipid–protein complexes (Pethica, 1958; Schulman *et al.,* 1955; Newton, 1960).

A destructive effect of several detergents on the membrane of *P. vulgaris* spheroplasts was reported by Nermut (1964). Sodium dodecylsulfate solubilizes *B. megaterium* membranes (Salton, 1957). Gilby and Few (1957, 1960) investigated the mode of action of detergents on *M. lysodeikticus* protoplasts. These authors think that the positively charged groups of cationic detergents react with the

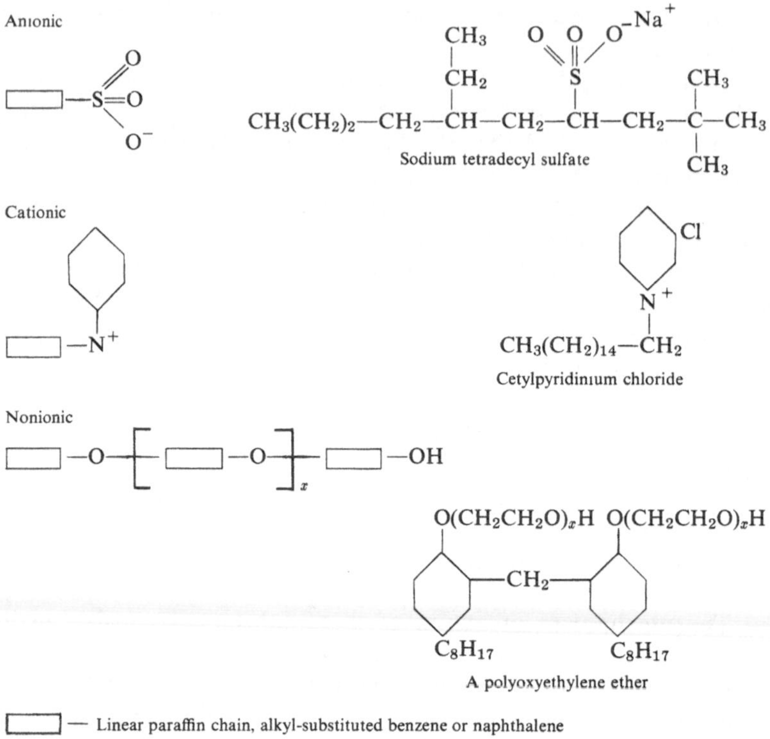

Anionic

$$CH_3(CH_2)_2-CH_2-CH-CH_2-CH-CH_2-C-CH_3$$

Sodium tetradecyl sulfate

Cationic

Cetylpyridinium chloride
$$CH_3(CH_2)_{14}-CH_2$$

Nonionic

$$O(CH_2CH_2O)_xH \quad O(CH_2CH_2O)_xH$$

$$-CH_2-$$

$$C_8H_{17} \qquad C_8H_{17}$$

A polyoxyethylene ether

— Linear paraffin chain, alkyl-substituted benzene or naphthalene

polyphosphatidic acids of the membrane. According to their data, lysis of the membranes by cationic detergents is a process secondary to disturbance of the membrane permeability. In contrast to cationic detergents, anionic detergents act on the protein component of the lipoprotein complex of the membranes. The nonionic detergents Triton, Tween 20, Tween 80, and Emasol also cause destruction of bacterial membranes and solubilization of the components, but their effect is milder since they rarely cause inactivating changes in enzyme proteins (Okui *et al.*, 1963; Dowben and Koeler, 1961). In some cases, nonionic detergents can even activate enzymes (Yonetani, 1959; Goldman *et al.*, 1963).

Cholate and desoxycholate compounds resemble detergents in their action. Desoxycholate and Tween 80 destroy the membranes of *M. lysodeikticus* (Lukoyanova *et al.*, 1961). The mechanism of the solubilizing effect of sodium desoxycholate on the lipoprotein complexes of membranes has been studied in the case of mitochondrial cytochromes. Desoxycholate displaces the lipid molecules from the cytochrome–lipid complex of the mitochondrial membrane (Wainio, 1960). Ball and Joel (1962) observed the solution of many components of the mitochondrial respiratory chain due to desoxycholate. Criddle *et al.* (1962) described the extraction of the structural protein of mitochondria by several detergents. The destruction of bacterial membranes by detergents and bile-acid salts indicates the important role of lipids in membrane formation.

Important information on the structure-forming role of lipids in biological membranes can be obtained from an investigation of the effect of lipase and phospholipase A, which, by destroying the integrity of neutral lipids and phospholipids, also lead to injury to the membrane structure. As is known, phospholipase A (lecithinase A) splits off the fatty acid bound with the β-hydrocarbon atom of phosphatidylcholine, but there is evidence that it also acts on diphosphatidylglycerol to form a lysoproduct (Van Deenen and De Haas, 1963; Marinetti, 1964). Lipase detaches fatty acids from glycerides. It has been shown that *B. megaterium* protoplasts in an isotonic medium are lysed by incubation with lipase (Spiegelman *et al.,* 1958; Vennes and Gerhardt, 1956). A *Mycoplasma* culture behaves in a similar way (Razin and Argaman, 1963).

The membranes of *M. lysodeikticus* contain phosphatidylglycerol, diphosphatidylglycerol, and a diglyceride—substrates for phospholipase A and lipase. In fact, *M. lysodeikticus* protoplasts in 1 *M* sucrose are lysed by the action of lipase and phospholipase A, as revealed by reduction of turbidity (Ostrovskii *et al.,* 1964; Lukoyanova, 1964). Changes in the ultrastructure of bacterial membranes due to lipolytic enzymes have been thoroughly investigated with the electron microscope by Biryuzova. The results will be given below. Here we will merely note that the structure of the bacterial membrane depends not only on the integrity of the phospholipids, as in the case of mitochondrial membranes, but on the integrity of glycerides as well. It is of interest that stable bimolecular lipid lamellae cannot be formed only from phospholipids or neutral lipids in model experiments. A definite ratio of the two components is essential (Thompson, 1946*b*). In lipoprotein complexes of bacterial membranes the proteins are screened by the lipids and are weakly attacked by proteolytic enzymes (Ostrovskii *et al.,* 1964; Razin and Argaman, 1963).

The role of lipids in bacterial membranes will be discussed in greater detail in Chapter III in connection with the localization and action of respiratory enzymes

STRUCTURAL SIMILARITIES OF BACTERIAL AND MITOCHONDRIAL MEMBRANES

Examination of the properties of model lipid systems indicates that there are two forms of lipid organization, lamellar and micellar, each of which can be converted to the other. However, the presence of protein and the whole range of physicochemical conditions in the cytoplasm make it difficult to extend these ideas unreservedly to biological membranes. Interpretation of the ultrastructure of biological membranes will require great effort and the development of new methods of investigation. The only approach at present to an understanding of the ultrastructure of biological membranes is electron microscopy of cells subjected to osmium or permanganate fixation and dehydration. These procedures, of course, distort the fine structure. Hence, until recently there has been no opportunity to picture the true organization of a particular biological membrane, particularly those specific differences which are due to the special features of the protein or lipid composition.

In 1943, Daoson and Danielli (1943) proposed a model for the structure of a biological membrane (Fig. 9), consisting of one or more bimolecular layers of lipids. Each polar surface of such a layer maintains a protein monolayer on itself. This model fits in well with electron-microscopic investigations of osmium- or permanganate-fixed membranes. Fixed specimens show that any biological membrane consists of three layers: protein–lipid–protein. The lipid is packed as two monomolecular layers. The two dense fine lines found in cell membranes fixed in this way are due to the deposition of os-

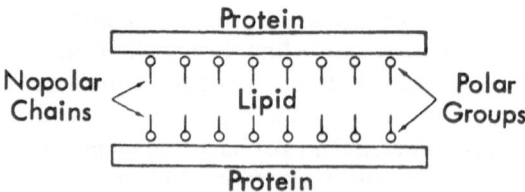

Fig. 9. Unit lipoprotein membrane (after Dawson and Danielli, 1953).

mium atoms at the protein–lipid boundary. The thickness of a typical lipid membrane is 75–80 Å. The bimolecular lipid layer accounts for about 50 Å, and the protein layers on each side are 10–15 Å thick (Stoeckenius, 1959; Stoeckenius *et al.*, 1960). Robertson suggested that such a membrane should be called a "unit membrane" (Robertson, 1959, 1961).

Different biological membranes consist of combinations and constructions of these unit membranes. In mitochondria the unit membrane is doubled. The outer envelope is also a unit membrane which forms transverse folds—the cristae. Hence, the mitochondrial membrane is 150 Å thick.

The structure of the cytoplasmic membrane and internal membranous structures in bacteria has still not been adequately studied, but it is obvious now that these structures are unit, and not double, membranes. As an example we can refer to *M. lysodeikticus* membranes obtained by lysis of the cell (see Fig. 7). The thickness of bacterial membranes, according to the data of different authors, varies within a narrow range (Table 5). The variation in the thickness

Table 5. Thickness of Membranes in Some Bacteria

Bacterium	Membrane thickness, Å	Author
Escherichia coli	60–80	Kellenberger and Ryter, 1958
Alcaligenes faecalis	80	Beer, 1960
Mycobacterium sp.	80	Koike and Takeya, 1961
Staphylococcus aureus	50	Suganuma, 1961
Rhodospirillum molischianum	80	Giesbrecht and Drews, 1962
Azotobacter chroococcum	70–80	Tchan *et al.*, 1962
Escherichia coli	80–120	Conti and Gettner, 1962
Escherichia coli	80	Ogura, 1963
Micrococcus lysodeikticus	75	Salton and Chapman, 1962
Pseudomonas sp.	75	Brown *et al.*, 1962
Rhizobium trifolii	80	Dart and Mercer, 1963
Acetobacter suboxydans	80	Claus and Roth, 1964
Listeria monocytogenes	80	Edwards and Stevens, 1963
Diplococcus pneumoniae	80–100	Tomasz *et al.*, 1963
Treponema microdentium	75–80	Listgarten *et al.*, 1963
Vibrio metchnikovii	75	Glauert *et al.*, 1963
Mycoplasma sp.	75–80	Domermuth *et al.*, 1964
Micrococcus roseus	75	Murray, 1963
Nitrosomonas europea	75	Murray, 1963
Nitrobacter agilis	75	Murray, 1963

of osmium- or permanganate-fixed bacterial membranes has not yet been fully explained.

New data indicate that the organization of the membranes is more complex than was previously supposed. These data also reveal some special features of mitochondrial and bacterial membranes. Recently obtained information regarding the chemical composition of membranes reveals a great diversity of lipid, protein, and carbohydrate composition. Factors such as the chain length of fatty acids and the molecular weight of protein and polysaccharides clearly will affect the membrane thickness. In view of this, the differences in the thickness of various unit membranes, which, according to modern information, can vary from 45 to 110 Å, are more understandable. Bacterial membranes differ slightly in thickness (see Table 5). This can be attributed to differences in the composition of the protein and lipid layers. For instance, osmium-fixed *E. coli* show two outer layers each 25 Å thick and an intermediate layer 30 Å thick. According to Claus and Roth (1964), for all gram-negative bacteria the thickness of the cytoplasmic membrane can be taken as 80 Å, that of each outer layer as about 30 Å, and that of the inner layer as about 20 Å. This structure of the cytoplasmic membrane seems to be peculiar to gram-negative bacteria.

These ideas of the common plan of structure and common parameters of bacterial, mitochondrial, and other biological membranes are based on numerous investigations of the structure of osmium- or permanganate-fixed cells.

However, examination of the literature on model lipid membranes clearly shows that the lamellar form of the membranes, which is always observed with such fixation, is not the sole form and may be replaced by the micellar type. Although this information has been obtained in model systems, it is a fascinating concept that the biological membrane is in fact not constructed from uniform lipid and protein monolayers, but has a more complex, mosaic structure, and that this structure is dynamic and subject to continuous change from one form to the other (Benson, 1964; Kavanau, 1965; Lucy, 1964; Green and Perdue, 1966). In this case, the nature of the lipoproteins, their interactions, and their localization in the membrane will ultimately determine the specific differences in membranes.

Although there is insufficient information on the structural specialization of individual parts of biological membranes, some experiments confirm the hypothesis that biological unit membranes which fulfill complex functions are better represented as a mosaic of lipoprotein complexes (Benson, 1964; Gent et al., 1964). These complexes may have the form of discrete lipid micelles separated from one another by protein. In addition, protein globules, representing the specialized enzymatic parts of the membrane, may be embedded in the aggregates of lipid micelles. New methods of cell fixation have revealed globular units 50–90 Å in diameter in some biological membranes (Sjöstrand, 1963a; Sjöstrand and Elfin, 1964; Cunningham et al., 1965). Globules of such size may be lipid micelles or globular proteins with enzymatic or hormonal properties.

There is also biochemical evidence of the mosiac nature of the membrane and its construction from different lipoproteins. Fragmentation of bacterial or mitochondrial membranes with desoxycholate causes not complete solubilization into molecules, but the appearance of lipoprotein complexes which differ in their enzymatic properties and composition (Green, 1962 a, b; Gel'man, 1963). The membranes of M. laidlawii are disintegrated by detergents to complexes with a sedimentation constant of 3.3S. If the detergent is removed in the presence of Mg^{++}, reconstruction of the membranes can be observed (Razin et al., 1965). Salton and Netschey (1965) observed fractionation of M. lysodeikticus membranes by detergents. The obtained lipoprotein complexes were homogeneous, but the sedimentation constant depended on the detergent used. Removal of Mg^{++} from M. lysodeikticus membranes splits off lipoprotein complexes of different size. By appropriate purification a complex with $Ks_{20} = 12.8S$ and an electrophoretic mobility of 1.54×10^{-4} cm^2 v^{-1} sec^{-1} in 0.005 Na-K phosphate buffer of pH 7.4 can be extracted. The complex contains malate dehydrogenase and lipid dehydrogenase. The number of basic and acid amino acids in the protein of the investigated complex is approximately the same. Hence, the pronounced acid properties of the complex may be due to additional acid groups, probably lipid phosphate groups. Nonpolar amino acids constitute about 55% (Zhukova et al., 1966).

MEMBRANE SUBUNITS

Whittaker (1964) regards the question of the "subunits" of cristae and internal membranes of mitochondria as one of the most intriguing problems of membrane structures. The "subunits" were found by negative-contrast staining with phosphotungstic acid, a method previously used to study the micellar form of lipids in model systems. Subunits were found for the first time on the surface of the cristae of beef heart mitochondria, and then on the cristae of mitochondria from various mammalian organs and the fungus *Neurospora crassa* (Fernandez-Móran, 1962; Parsons, 1963a, b; Stoeckenius, 1963). The subunits were mushroom-shaped and consisted of a head 80–100 Å in diameter and a cylindrical stalk 50 Å long and 40 Å broad embedded in the membrane. Stoeckenius calculated the number of particles as about 3000 per μ^2. D. Smith (1963) observed dumbbell-shaped subunits in the mitochondria from blowfly flight muscles.

A discussion at a symposium of the Sixth Biochemical Congress showed that most scientists do not regard the subunits as fixation artifacts (Slater, 1964). It is of interest that these subunits are not found either in the endoplasmic reticulum or in erythrocytes (Stoeckenius, 1963). They are found only on the cristae and on the inner side of the mitochondrial membrane, i.e., on the membranes carrying the respiratory chain. They were recently found in the mitochondria of the yeast *Endomyces magnusii* and in the mitochondria of higher plants (Meisel' *et al.,* 1964; Biryuzova and Meisel', 1964; Nadakavukaren, 1964).

In view of the functional similarity of mitochondria and bacterial membranes, it was to be expected that their ultrastructure would show some common features. We collaborated with V. I. Biryuzova to investigate isolated membranes of *M. lysodeikticus* (Biryuzova *et al.,* 1964). We used negative-contrast staining (Parsons, 1963a, b). The bacteria were lysed with lysozyme in a medium consisting of 10^{-3} M K-Na phosphate buffer (pH 7.4) and 10^{-4} M MgSO$_4$. The membranes were separated at 30,000 g and washed twice.

On the surface of the membranes we found mushroom-shaped subunits similar to those on the mitochondrial cristae of higher organisms. The diameter of the spherical head of the subunits of *M. lysodeikticus* membranes was approximately 80 Å and the width of the stalk was 25–30 Å. The relatively fine membranes carrying the subunits lay inside the cytoplasmic membrane, on which such structures could not be detected (Fig. 10).

Subunits have also been found on the internal membranes of *Eubacterium sp.,* on the cytoplasmic membrane and internal membranes of *Bacillus stearothermophilus* (Bladen *et al.,* 1964; Abram, 1965), and also on the membranes of *R. rubrum* (Löw and Afzelius, 1964).

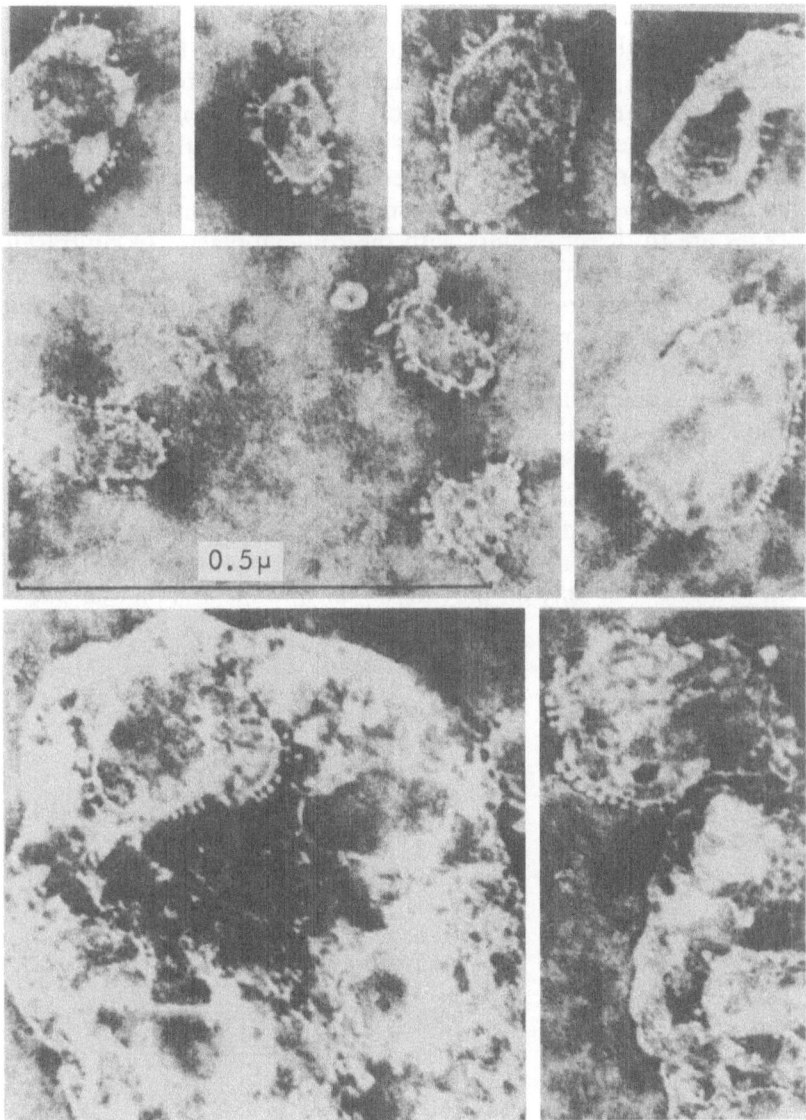

Fig. 10. Subunits of *Micrococcus lysodeikticus* membranes.

The discovery of subunits on bacterial membranes indicates that the ultrastructure of these membranes is similar to that of mitochondrial membranes of cells very far removed from the evolutionary standpoint, *viz.*, animal cells. Nevertheless, mitochondrial and bacterial membranes are indisputably different in morphology. In bacteria, the membranes consist of various invaginations of the cytoplasmic membrane and are not separated from the cytoplasm. The subunits situated on the cytoplasmic membrane and internal membranes are in direct contact with the cell cytoplasm. In contrast to this, the mitochondrial cristae are invaginations of the inner layer of the double mitochondrial membrane. The outer layer isolates the mitochondria, separating it from the other structural elements and the cytoplasm. The cristae and, hence, the mushroom-shaped subunits are embedded in the mitochondrial matrix, which is isolated from the cell cytoplasm.

Figure 10 shows that the subunits are situated only on one side of *M. lysodeikticus* membranes. This agrees with Abram's observations of membranes from *B. stearothermophilus*. According to Parsons (1963a, b), the outer mitochondrial membrane lacks these particles. They can be seen only on the internal membrane and on one side of the cristal membrane, the side turned towards the mitochondrial

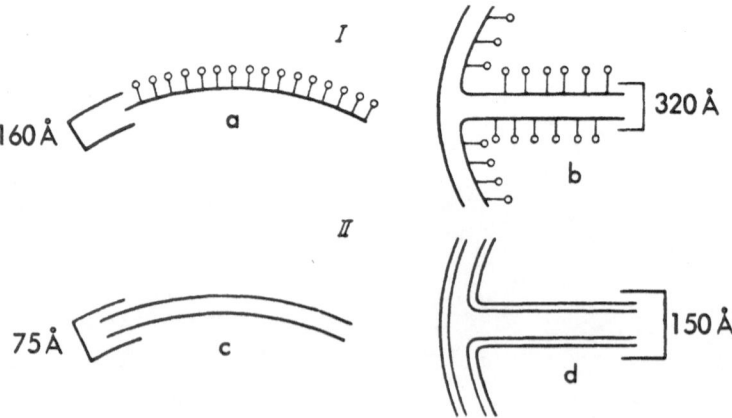

Fig. 11. Relative dimensions of bacterial and mitochondrial membranes after fixation by various methods. I — Negative staining with PTA: (*a*) membranes of *Micrococcus lysodeikticus;* (*b*) mitochondrial membrane. II — Contrast staining with 1% OsO$_4$: (*c*) membranes of *Micrococcus lysodeikticus;* (*d*) mitochondrial membranes.

matrix (Fig. 11). The unilateral disposition of the subunits confirms the view of a geometrically asymmetric unit membrane in biological structures (Sjöstrand, 1963b; Green, 1964a, b).

The discovery of subunits on structures responsible for electron transport suggests their connection with this process. The function of the subunits is discussed in Chapter III. In this chapter, we will confine ourselves to consideration of the structural interrelationships of the subunits and membrane. The investigation of the composition, structure, and mode of attachment of the subunits to the membrane requires the separation of the subunits from it. This is a very difficult problem, since disappearance of the subunits from the membrane can be effected, but the membrane itself is usually fragmented, and then it is very difficult to separate the mixture of particles. This explains some of the contradictory information on the composition and mode of attachment of the subunits to the membrane.

According to Green's school, treatment of mitochondrial membranes with ultrasound and then with cholate or desoxycholate causes detachment of the subunits, which can then be isolated as a pure fraction (Blair et al., 1963). However, other authors have failed to achieve detachment of the subunits by ultrasound (Greville et al., 1964). Treatment with bile acid salts is also open to question. Lusena and Dass (1963) found that if mitochondrial fragments are incubated longer than 8 min with desoxycholate, the subunits will disappear altogether.

The subunits of mitochondria are apparently aggregates of lipoprotein complexes (Blair et al., 1963), although this will not be demonstrated conclusively until an absolutely pure fraction of subunits is obtained, without any admixture of carrier membrane elements. If Green (1964b) is right, however, in his hypothesis that the subunits are dumbbell-shaped and that one end of each dumbbell contacts its neighbor, thus forming a continuous "carrier" membrane, separation of the subunits would be associated with disintegration of the whole membrane and there would be no reason to expect any differences in the composition of the carrier membrane and the subunits.

According to our results, however, the situation is not so simple. As already mentioned, the action of pancreatic lipase or phospholipase A on M. lysodeikticus membranes leads to a reduction in the turbidity of the protoplasts. This suggests injury to the cytoplasmic membrane. It was thus of interest to determine the nature of the

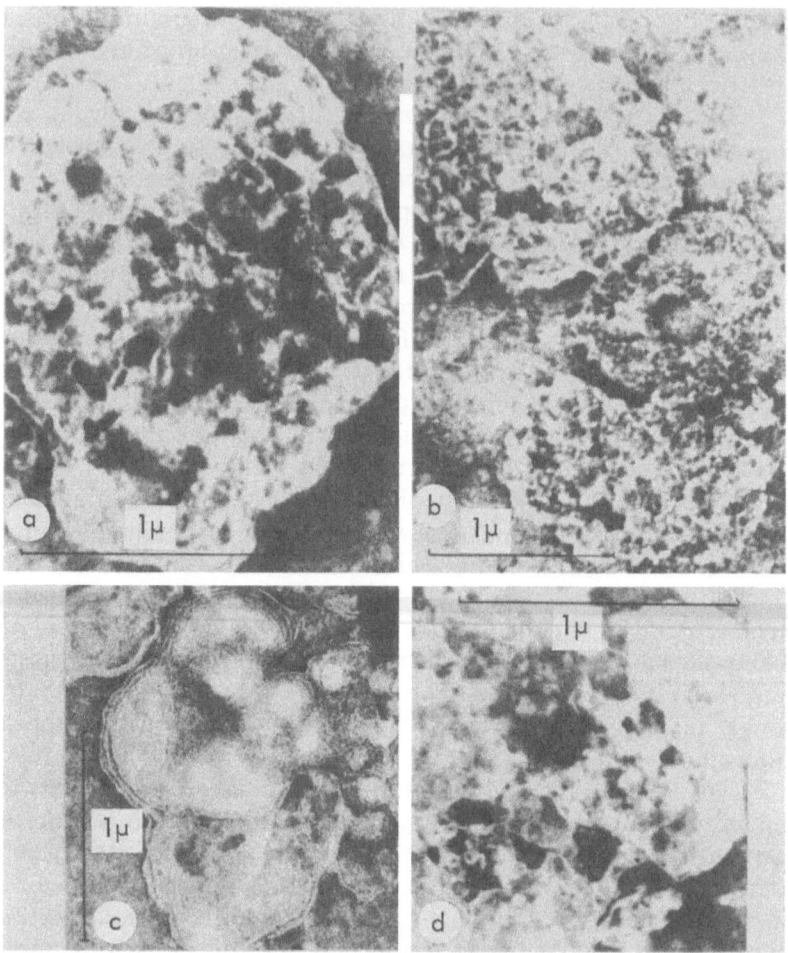

Fig. 12. Membranes after lipase treatment.

injury to the subunits and their condition. Preparations of membranes before and after treatment with lipolytic enzymes were investigated by negative-contrast staining (Brenner and Horne, 1959; Parsons, 1963a, b).

Unlike detergents, lipolytic enzymes do not cause solution of the membranous structures as such (Lukoyanova et al., 1961), but the subunits usually disappear after treatment of the membranes with lipase (Fig. 12a). They can be observed only in a very few cases. In-

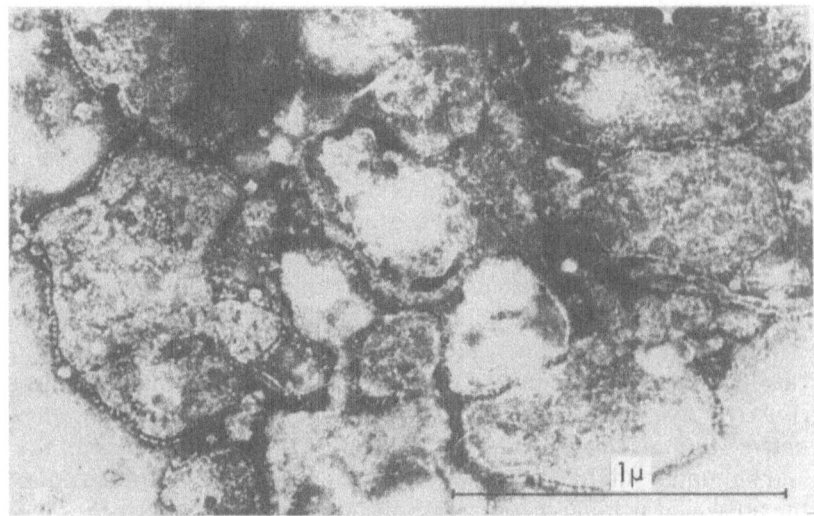

Fig. 13. Membranes before treatment with lipolytic enzymes.

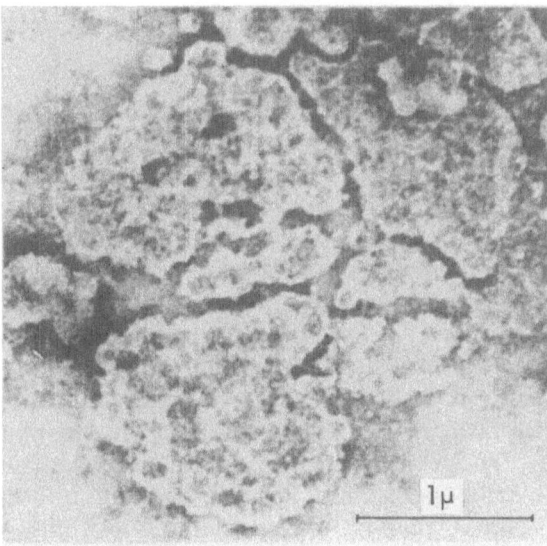

Fig. 14. Membranes after treatment with phospholipase A.

jured membranes with a perforated surface are seen in the field of vision (Fig. 12b). Stratification of the membranes is frequently observed. This is particularly distinct on the outer limiting membrane (Fig. 12c). Sometimes the outer membrane disappears, whereas the internal membranes remain and appear compact, their connection with one another being unaffected (Fig. 12d). The outer bounding membrane enclosing several internal membranes is seen on the control preparations (Fig. 13). Sometimes the internal membranes are seen separately. Mushroom-shaped subunits 60–80 Å in diameter can be seen on the surface of the internal membranous structures.

Treatment with phospholipase A produces similar pictures, but stratification of the membranes is not observed (Fig. 14). As in the case of lipase treatment, no mushroom-shaped subunits are found (Lukoyanova and Biryuzova, 1965). It is obvious that lipids in their native state are necessary to ultrastructure of bacterial membranes and subunits. Slight but irreversible damage follows detachment of the fatty acid from glycerides and phospholipids and the appearance of lysoproducts. Lucy and Glauert (1964) and Bangham and Horne (1964) demonstrated the extreme lability of phospholipid membranes in model systems. The addition of lysoproducts to lamellae will cause their transition to the micellar state and eventually perforation of the membranes.

The subunits disappear fairly rapidly, and the membrane does not disintegrate. The reasons for the relative stability of the membrane to the action of lipolytic enzymes and the rapid disappearance of the subunits are obscure. Some information obtained from an analysis of the lipids in the membranes before and after their treatment with lipolytic enzymes will be given in the next chapter.

In considering the methods of structural organization of membranes, we started with the assumption of the importance of the lipid and protein composition. This approach is probably not wholly justified. In fact, the membrane of M. lysodeikticus has subunits and "outwardly" does not differ in any way from a mitochondrial membrane. Yet the differences in lipid composition, as we saw above, are immense. This suggests particular physicochemical relationships between the components are a necessary condition in the structure of membranes and subunits, and that identity of composition is not essential.

Chapter III

The Respiratory Chain of Bacteria

GENERALIZED SCHEME OF RESPIRATORY CHAIN OF BACTERIA

The vast majority of bacteria have a respiratory chain,* i.e., a system of spatially organized dehydrogenases and cytochromes which bring about electron transport.† The difference between aerobic bacteria and facultative anaerobes lies mainly in the terminal electron acceptor. In aerobes the acceptor is oxygen, and in facultative anaerobes it can be either oxygen or an oxidized acceptor of the nitrate or sulfate type. Butyric acid and lactic acid bacteria form a special group as they have no cytochromes. Hydrogen transfer in these bacteria is effected in a donor–acceptor system and phosphorylation takes place at the substrate level (Dolin, 1963b). Despite the absence of a specialized system of respiratory-chain enzymes, these bacteria can absorb some oxygen from the environment by means of flavin oxidases. In obligate anaerobes (butyric acid bacteria), this process usually leads to death of the cell, whereas facultative anaerobic bacteria such as lactic acid bacteria can tolerate the presence of oxygen.

Photosynthetic bacteria occupy a special place among the obligate anaerobes. Some representatives of this group can develop a normal respiratory chain when grown in aerobic conditions in the dark.

There are several reviews devoted to electron-transport enzymes

---

* Respiratory chain is a synonym of "electron transport chain."
† In this book we use the most widely adopted term, "electron transport," although there is not yet sufficient proof that the transport of reducing equivalents from substrate to oxygen in the respiratory chain is effected only in the form of electrons (Dixon and Webb, 1961).

in bacteria (Dolin, 1963a, b; Smith, 1963; Newton and Kamen, 1963). We will attempt to give a very general description of the respiratory chain in bacteria, and then we will discuss in more detail the data relating to the properties and localization of particular enzymes in the cell.

A striking feature of the bacterial respiratory chain is the diversity of qualitative and quantitative enzyme composition. Hence, the construction of a generalized scheme presents considerable difficulties We have made such an attempt in order to facilitate the comparison of the bacterial respiratory chain with the mitochondrial respiratory chain, but we do not claim to cover all the material.

The respiratory chain of aerobic bacteria contains several dehydrogenases, ubiquinones or naphthoquinones, and cytochromes of groups A, B, and C.

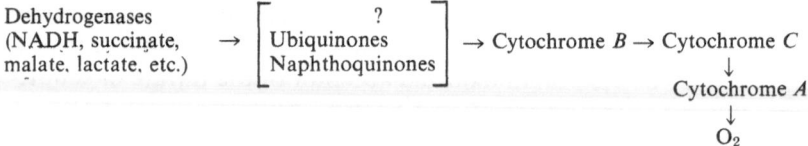

$$\begin{array}{l}\text{Dehydrogenases} \\ \text{(NADH, succinate,} \\ \text{malate, lactate, etc.)} \end{array} \rightarrow \left[\begin{array}{c} ? \\ \text{Ubiquinones} \\ \text{Naphthoquinones} \end{array}\right] \rightarrow \text{Cytochrome } B \rightarrow \text{Cytochrome } C \\ \downarrow \\ \text{Cytochrome } A \\ \downarrow \\ O_2$$

Respiratory chain of aerobic bacteria

As this scheme shows, electrons can enter the respiratory chain through various dehydrogenases, which are firmly bound with the whole enzyme complex of the respiratory chain. Thus, in bacteria, the first stage in the oxidation of substrates is not necessarily their dehydrogenation by cytoplasmic enzymes to produce reduced nicotinamide adenine dinucleotide (NADH); the process can begin with flavin enzymes which transfer an electron directly to the respiratory chain. In many chemosynthetic bacteria, the enzymes responsible for the initial oxidation of the inorganic substrate (hydroxylamine, Fe^{++}, nitrite, H_2, etc.) are also directly attached to the respiratory chain (Doman and Tikhonova, 1965). In this respect, the bacterial respiratory chain differs considerably from the mitochondrial respiratory chain, where electrons enter the chain mainly from two firmly bound dehydrogenases—succinic dehydrogenase and NADH dehydrogenase.

The prosthetic groups of some dehydrogenases of the bacterial respiratory chain have been investigated, and it has been found that they are, as in mitochondria, flavin nucleotides. Flavoprotein enzymes transfer electrons to the cytochrome system. There are data which indicate that the bacterial respiratory chain, like the mitochondrial respiratory chain, includes ubiquinones and naphthoquinones, the site of which in the respiratory chain will be discussed below. Cytochromes of groups A, B, and C are present in the bacterial respiratory chain. Despite their similar functions, they differ from mitochondrial cytochromes in several properties. The cytochromes and terminal oxidases in aerobic bacteria are characterized by great diversity, as shown below (after Smith, 1963).

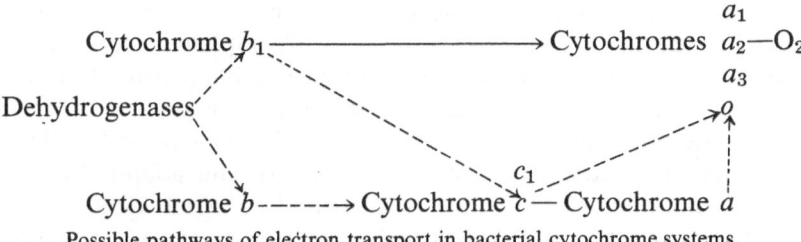

Possible pathways of electron transport in bacterial cytochrome systems.

In the majority of facultative anaerobes, the respiratory chain is similar in its set of components to that of aerobic bacteria, but the terminal electron acceptor can be various oxidants, such as nitrates, sulfates, and oxidants produced in the process of nitrogen fixation, although oxygen can also be used. In these variants the electron-transport scheme can be represented as follows (Ishimoto and Egami, 1957; Dolin, 1963a):

$$\text{Substrate} \rightarrow \text{Dehydrogenase} \rightarrow \text{Cytochromes} \quad \begin{bmatrix} \text{Nitrate reductase} \rightarrow NO_3^- \\ \text{Nitrite reductase} \rightarrow NO_2^- \\ \text{Nitric oxide reductase} \rightarrow NO \\ \text{Sulfate reductase} \rightarrow SO_4^{-2} \\ \text{Thiosulfate reductase} \rightarrow S_2O_3^{-2} \\ \text{etc.} \end{bmatrix}$$

The above material indicates the very diverse composition of bacterial electron transport systems, which may reflect the course of the evolutionary process. According to Oparin (1957, 1960), the first organisms were anaerobes and heterotrophs. These were followed by photosynthetic forms and, finally, by aerobic forms, autotrophic and heterotrophic. These views on the evolution from anaerobic to aerobic forms have been developed by many authors (Krasnovskii, 1957; Horecker, 1962; Imshenetskii, 1962; Gaffron, 1964; Lipmann, 1966; Yamanaka, 1964). The indicated evolutionary series can be traced to some extent by an analysis of electron-transport mechanisms. Obligate anaerobic bacteria lack cytochromes and ubiquinones and use soluble flavin enzymes for electron (hydrogen) transport in a substrate → substrate chain (Dolin, 1963b; Oparin et al., 1964). Next, of course, come the facultative anaerobes, which have acquired cytochrome components, but some of them still lack terminal oxidases for transferring the electron to oxygen, and the final electron acceptor consists of inorganic oxidants. The typical respiratory chain peculiar to obligate aerobes has a complete set of essential components, as in cells of higher organisms, but differs from the latter in the fairly considerable variation in the concentration and even the composition of components of the chain in the same organism. This is clearly connected with the great plasticity and adaptibility observed among microorganisms (Kluyver and van Niel, 1959).

The functional mechanism of the respiratory-enzyme chain in bacteria has not been as intensively studied as that of the respiratory chain in the mitochondria of higher organisms. This is obviously due to the tacit acceptance of the identity of these mechanisms by investigators, although, generally speaking, this identity has not been proven. Hence, some information on electron transport via the respiratory chain in mitochondria is in order.

The question of the electron-transport mechanism has two aspects. First of all, what changes take place in the electron carrier, i.e., does the transfer involve two electrons or one, and are free radicals formed, or is the transport of an ionic nature? Secondly, how is an electron transferred from carrier to carrier, and why are the carriers structurally organized as complexes in a lipoprotein membrane?

As regards the first aspect, it is now accepted that electron transfer from the substrate to flavoprotein is effected in the form of two reducing units at once (electron + hydrogen atom), but that electrons are transferred alone through the cytochromes to oxygen. Lehninger (1964), however, thinks that there may be other modes of electron transport: (1) the electrons are transferred singly through the whole chain, or (2) the electrons are transferred in twos through the whole chain, and paired cytochrome molecules may be used for such transport:

$$\text{Substrate} \xrightarrow{3} \text{NAD} \xrightarrow{3} \text{FP} \mid \begin{array}{c} b \to c \to a_1 \\ \nearrow \qquad\qquad \searrow \\ \qquad\qquad\qquad O_2 \\ \searrow \qquad\qquad \nearrow \\ b \to c_1 \to a_3 \end{array}$$

There are at least four viewpoints regarding the interaction of the carriers in the lipoprotein membrane. In the opinion of Green and his colleagues, electron transport between the individual respiratory-chain enzymes forming lipoprotein complexes is effected by low-molecular substances of the ubiquinone or nonheme iron type. Chance puts forward several hypotheses on this question: (1) electron transport may be effected by rotation of the whole enzyme molecule or part of it around its axis so that it comes into contact in turn with two "neighbors" in the respiratory chain; (2) electrons may be transferred between enzymes by a system of π-electrons; and (3) electrons may be transferred by the collision of enzymes, which are pictured as pendulum bobs suspended from the membrane.

Model experiments have recently shown that the semiconducting and optical properties of cytochrome c change when it is complexed with ubiquinone. This is regarded as evidence of the presence of a semiconducting mechanism in electron transport in the respiratory chain (Snart, 1964). For a fuller description of the mechanism of electron transport, we refer the reader to reviews on the subject (Lehninger, 1959, 1964; Pullman and Pullman, 1965; Green, 1964a, b; Ernster and Lee, 1964; Chance and Williams, 1956; Chance, 1964; Green and Fleischer, 1964).

DEHYDROGENASES

Bacterial cells contain numerous dehydrogenases, which take part in anaerobic transformations of substances (glycolysis, various fermentations), in the tricarboxylic acid cycle, and in other processes leading to the generation of NADH or NADPH. The majority of these enzymes are dissolved in the cytoplasm of the bacterial cell and are connected with the respiratory chain only through NAD, the universal hydrogen carrier. The oxidation of NADH involves the flavin enzyme NADH dehydrogenase, which is insoluble and firmly bound with the respiratory chain. The other enzyme, succinic dehydrogenase, oxidizes succinate, but the strength of its connection with the respiratory chain differs in different bacteria.

In addition to NADH dehydrogenase and succinic dehydrogenase, bacteria have numerous dehydrogenases which are firmly bound in the respiratory chain system (lactic dehydrogenase, glucose dehydrogenase, glucose-6-phosphate dehydrogenase, lactose dehydrogenase, malic dehydrogenase, alcohol dehydrogenase). The prosthetic groups of these enzymes have hardly been studied at all, and the presence of flavin nucleotides has been shown only in certain cases.

A list of the firmly bound dehydrogenases in the bacterial respiratory chain is given in the section on the localization of respiratory-chain enzymes. Here we will merely describe the properties of the individual enzymes of this group for which there are data regarding the mode of purification, nature of the prosthetic group, and effect of inhibitors. NADH dehydrogenase, succinic dehydrogenase, and dehydrogenases firmly bound with the respiratory chain are difficult to solubilize and to purify. This may be the reason why these enzymes have been so little investigated. Difficulty in solubilizing membrane-bound dehydrogenases in *E. coli* and *Salmonella typhosa* has been reported. Treatment with desoxycholate, butanol, digitonin, or ultrasound led to extraction of the proteins into the aqueous phase, but this inactivated the succinic and lactic dehydrogenases (Kidwai and Murti, 1965).

A characteristic feature of many bacteria is the simultaneous presence of two dehydrogenases which effect dehydrogenation of the same substrate. One of them is soluble and requires NAD for its action; the other is firmly bound in the respiratory chain and does not require NAD. The malic dehydrogenases of *M. lysodeikticus* provide a good example of this (Cohn, 1956, 1958; Gel'man *et al.*, 1960*a*; Gel'man, Zhukova, Lukoyanova, and Oparin, 1959 and 1963). The malic and lactic dehydrogenases from *A. vinelandii* do not require pyridine nucleotides for their action (Jones and Redfearn, 1966).

An interesting relationship between the soluble and particulate malic dehydrogenases has been discovered in different species of *Pseudomonas*. The cells of most species contain only the soluble malic dehydrogenase, which requires NAD for its action. *Pseudomonas* species lacking the soluble enzyme contain a malic dehydrogenase firmly bound with the respiratory chain which does not require NAD (Kornberg and Phizackerley, 1961). The hypothesis of a flavin pros- thetic group in the enzyme is supported by data indicating that flavin nucleotides are present in membrane preparations and that the en- zyme effectively reduces typical acceptors of flavin dehydrogenases (phenazine methosulfate, 2,6-dichlorophenolindophenol) and $2 \times 10^{-3} M$ amobarbital suppresses its activity by 80% (Francis et al., 1963). Malic dehydrogenase has been purified 100-fold by treatment of a particle suspension with ultrasound, centrifugation at 100,000 g, salting out, and fractionation on DEAE-cellulose. An essential con- dition for the activity of the enzyme is the addition of membrane phospholipids, FAD, and 2-methyl-1,4,-naphthoquinone (Francis and Phizackerley, 1965).

A malic dehydrogenase firmly bound with the respiratory chain is found in *Mycobacterium avium* (Kimura and Tobari, 1963). For extraction of the enzyme, the particulate cell fraction was first treated with acetone, and then with Tris buffer and ammonium sulfate. The obtained preparation had high activity but was not homogeneous. The enzyme was contained in a lipoprotein complex which precipi- tated at 77,000 g. The complex could not be degraded with phospho- lipase A, or lipase, or detergents. The best acceptor for this enzyme is phenazine methosulfate. Its activity is increased by $2.2 \times 10^{-6} M$ FAD, which is indicative of a flavin prosthetic group. Tobari suc- ceeded in extracting an apomalic dehydrogenase from acetone pow- der and purifying it by salting out and fractionation on DEAE-cellu- lose. The activity was manifested in the presence of FAD and lipid from *M. avium* cells. The main active principle in the lipid was prob- ably diphosphatidylglycerol (Tobari, 1964).

According to Cohn (1956, 1958), malic dehydrogenase is firmly bound with *M. lysodeikticus* membranes and becomes soluble when the membranes are treated with Versene. Using Cohn's method, we extracted an enzyme preparation which contained NADH dehydro- genase in addition to malic dehydrogenase from *M. lysodeikticus* membranes. By salting out, followed by dialysis, centrifugation at 160,000 g for 15 min, and fractionation of the supernatant fluid on DEAE-cellulose, we managed to purify the preparation 10- to 12-fold (Gel'man, Zhukova and Oparin, 1963). The preparation contained lipids, including carotenoids. The infrared spectrum of the lipid ex- tract was identical with that of the membrane lipids (Oparin et al., 1963). Chromatography revealed the same phospholipids which

Macfarlane (1961*a, b*) identified in membranes. No pyridine nucleo-tides were found in the preparation, but there were flavin nucleotides, which are possibly the prosthetic groups of both enzymes. That lipid is essential for the activity of both enzymes, particularly malic dehy-drogenase, was shown by the action of lipase or phospholipase A on the preparation (Oparin, Gel'man and Zhukova, 1965). These results agree with those of Tobari, who showed that lipid is essential for the activity of malic dehydrogenase from *M. avium* membranes. Hauge and Hallberg (1964) reported inactivation of glucose dehydrogenase when membranes of *B. anitratum* were digested with phospholipase A. The interaction of the dehydrogenase with the lipids of the mem-brane carrying the respiratory chain apparently promotes the con-formation necessary for the activity of the enzyme.

An alcohol dehydrogenase which does not require NAD has been found in the cells of *Acetobacter peroxydans* (Nakayama, 1961). The enzyme was purified on aluminum-oxide columns and by starch-gel electrophoresis. Cytochrome c_{553} was identified as the prosthetic group of the enzyme. Flavin nucleotide was probably present too. In this case the enzyme is similar to yeast succinic dehydrogenase and lactic dehydrogenase. *Gluconobacter suboxydans* membranes are ex-ceptionally rich in firmly bound dehydrogenases (De Ley and Kersters, 1964; Kersters *et al.*, 1965). The substrates of these dehy-drogenases are primary and secondary alcohols, poly alcohols, pentoses, and hexoses. The enzymes can be solubilized with difficulty by treatment of the membranes with Triton X-100 and Versene at pH 7.6. The addition of NAD, NADP, and flavin nucleotides has no effect on the activity of the dehydrogenases. The dehydrogenases pass into the soluble fraction in the form of a complex with cytochrome c_{553}. According to Iwasaki (1960), the lactic dehydrogenase of *A. sub-oxydans* has flavin nucleotide as a prosthetic group and is firmly bound with cytochrome *b*.

A D-allohydroxyproline dehydrogenase bound in particles (100,000 g) of *Pseudomonas striata* can be extracted with Versene and is characterized by selectivity towards acceptors. The enzyme does not interact with methylene blue, $K_3Fe(CN)_6$, or pyocyanine, and re-duces only phenazine methosulfate. The addition of NAD, FAD, or FMN has no effect on the enzyme activity (Adams and Newberry, 1961).

The respiratory chain of *A. suboxydans, Pseudomonas fluorescens,* and *B. anitratum* has a firmly bound glucose dehydrogenase (Hauge, 1961*a, b*). The activity of the enzyme does not depend on added NAD or NADP. The enzyme is found both in the heavy fraction (18,000 g, 90 min) and in the light fraction (240,000 g, 120 min). The enzyme can be solubilized with desoxycholate from the particulate fraction of *B. anitratum* and purified almost to homogeneity. The nature of the prosthetic group has not been determined. The reduced enzyme shows a maximum at 337 mμ (Hauge, 1964). A partially purified aldose dehydrogenase has been obtained from the cell membranes of the halophilic bacterium *Pseudomonas* No. 101 (Maeno, 1965). The particles were extracted with a mixture of desoxycholate and cholate. The soluble fraction was separated at 105,000 g and the proteins were fractionated by salting out and on Sephadex G-200. The prosthetic group of the enzyme has not been determined.

A lactose dehydrogenase from *Pseudomonas graveolens* has been purified. The particulate fraction (40,000 g) obtained after disruption of the cells with ultrasound was treated with a mixture of desoxycholate and *n*-butanol for solubilization of the enzyme. Then the enzyme was purified 40- to 70-fold by treatment with chloroform and protamine sulfate, salting out with ammonium sulfate, and fractionation with ethanol. The prosthetic group is probably FAD. The purified enzyme interacts with 2,6-dichlorophenolindophenol, methylene blue, and $K_3Fe(CN)_6$(Nishizuka *et al.,* 1960; Nishizuka and Hayashi, 1962).

A complex of formic dehydrogenase and cytochrome b_1 has been obtained from *E. coli* particles. Desoxycholate was used to extract the enzyme. After purification the activity of the preparation was increased 20-fold. The preparation contained 0.5 mμ mole of flavin, 1.5 mμ mole of cytochrome b_1 per mg of protein, and 3% of lipid. The interaction of the enzymes is not affected, since cytochrome b_1 is completely reduced by formic dehydrogenase in the presence of substrate (Linnane and Wrigley, 1963).

In addition to the firmly bound dehydrogenases, which do not require NAD to act, there are weakly bound types, such as the malate:vitamin K reductase from *M. phlei*. The prosthetic group of the enzyme is apparently FAD (Adelson *et al.,* 1964; Asano and Brodie, 1963). The enzyme requires a small amount of phospholipids for its activity (Asano *et al.,* 1965). Weakly bound NADH dehydrogenases have also been described from *M. tuberculosis* (Heinen *et al.,* 1964).

The main enzymes of the initial stage in the respiratory chain, NADH dehydrogenase and succinic dehydrogenase, have received little study. It has been found that the succinic dehydrogenase of

Propionibacterium pentosaceum becomes soluble only after treatment with butanol. The purified enzyme preparation contains cytochrome *b*, in addition to flavin (Singer and Lara, 1958). A highly purified succinic dehydrogenase has been obtained from *M. lactilyticus* cells. The enzyme resembles mitochondrial succinic dehydrogenase, since it is an iron flavoprotein (Warringa and Guiditta, 1958; Warringa *et al.*, 1958). A menadione reductase capable of dehydrogenating NADH and reducing menadione (vitamin K_3), O_2, and triphenyltetrazolium chloride has been extracted from *E. coli* particles by solubilization with desoxycholate. The last two reactions occur only in the presence of menadione. The enzyme was purified 110-fold on Sephadex G-200 and calcium phosphate gel (the activity was measured by tetrazolium assay). The prosthetic group is probably FAD, the addition of which stimulated the enzyme activity by 25–50% (Bragg, 1965*a*).

Chemoautotrophs have specific enzymes which effect the oxidation of inorganic substrates. Owing to the difficulties in work with this group of bacteria, only a few enzymes have been isolated and purified. The results of investigations of the hydrogenase of hydrogen bacteria are given in several papers (Bone, 1963; Bone *et al.*, 1963; Repaske, 1961; Repaske, 1962). It is believed that the acceptance of hydrogen by hydrogenase involves the reversible breakage and hydrogenation of metal sulfide bonds (Bone, 1963).

It has been established that the hydrogenase of *Desulfovibrio* is an iron-containing protein which does not require other cofactors (Sadana and Morey, 1960; Riklis and Rittenberg, 1961). The enzyme, purified 1300-fold, had a molecular weight of about 10,000 and an oxidation–reduction potential of -0.400 v.

UBIQUINONES AND NAPHTHOQUINONES

Naphthoquinones and ubiquinones are found in bacterial cells. Ubiquinones are derivatives of 2.3-dimethoxy-5-methylbenzoquinone which have a polyisoprenoid chain of varying length in the sixth position.

$$CH_3O-\overset{3}{\underset{2}{\bigcirc}}-\overset{5}{\underset{6}{}}-CH_3 \quad CH_2CH=\overset{CH_3}{\underset{}{C}}-CH_2)_nH$$

The following natural ubiquinones are known; (UQ_{10}), $n = 10$; (UQ_9), $n = 9$; (UQ_8), $n = 8$; (UQ_7), $n = 7$; (UQ_6), $n = 6$.

Vitamins K (menaquinones) are regarded as menadione derivatives which differ from menadione in having a radical in the third position of the naphthoquinone ring. In the case of menadione, H replaces the radical.

2-Methyl-1,4-naphthoquinone, menadione (vitamin K_3)

The compound having the phytol-C20 residue with one double bond in the third position is designated vitamin $K_{1(20)}$, phylloquinone (Wagner and Folkers, 1964).

2-Methyl-3-phytyl-1,4-naphthoquinone (vitamin K_1)

Vitamins of the K_2 group have a side chain with a varying number of isoprenoid units, but usually contain one double bond to each C5-structural unit. The length of the side chain is indicated after the name of the quinone, e.g., $K_{2(35)}$. This vitamin is represented below (Wagner and Folkers, 1964):

2-Methyl-3-farnesylgeranylgeranyl-1,4-naphthoquinone

Three naphthoquinones related to vitamin K_2 are contained in the cells of *Hemophilus parainfluenzae*. They differ from vitamin K_2 in having the methyl group in the second position replaced by hydrogen. Hence, these compounds are called 2-demethyl vitamins K_2. The number of carbon atoms in the isoprenoid chain is 25, 30, and 35 (Lester *et al.*, 1964). A naphthoquinone containing eight isoprenoid units in the side chain, but with seven double bonds, has been isolated from *C. diphtheriae* (Scholes and King, 1965*b*). The authors call this compound MK-8 (2H).

Table 6 gives information on the occurrence and amount of quinones in investigated bacteria. As these data show, gram-positive bacteria do not contain ubiquinones. Ubiquinones predominate in gram-negative bacteria, but the latter may contain both ubiquinones and naphthoquinones (mainly vitamin K_2). Cells of *E. coli* and *P. vulgaris* contain both ubiquinones and naphthoquinones.

To clarify the function of ubiquinones and naphthoquinones in bacteria, analysis of their occurrence in relation to aeration conditions during cultivation is in order.

The naphthoquinone content of some gram-positive bacteria depends on the cultivation conditions. In anaerobically grown *S. albus,* for instance, the naphthoquinone content is much lower than that of cells grown aerobically (see Table 6). It has been shown for *E. coli* that aerobic conditions favor synthesis of ubiquinones, whereas biosynthesis of vitamin K predominates under anaerobic conditions (Polglase *et al.*, 1966). However, no differences in the ubiquinone or naphthoquinone content of cells grown in aerobic or anaerobic conditions have been found in the case of some gram-negative bacteria (*P. vulgaris, Pseudomonas pyocyanea*). Although the quinone content of aerobic and facultative anaerobic bacteria is not necessarily related directly to the oxygen content of the medium, obligate anaerobic bacteria (*Clostridium sporogenes, L. casei*) usually lack both quinones and cytochromes. This indicates that ubiquinones and naphthoquinones are definitely related to the nature of the oxidative metabolism of bacteria. Some authors attribute the absence of naphthoquinones in bacterial spores to the weak oxidative metabolism.

The group of obligate anaerobic intestinal bacteria is a special case; many of these bacteria have been isolated from the rumen of mammals. Some retain certain respiratory-chain components (cytochromes *b* and *c*), but oxygen is never a terminal electron acceptor.

Table 6. Ubiquinones and Naphthoquinones in Bacterial Cells

Organism	Cultivation conditions	Ubiquinone	Naphtho-quinones	Author
		μmole/g dry cell weight		
Gram-positive				
Bacillus subtilis	Aerobic	<0.001	0.7	
Bacillus subtilis	Spores, aerobic	<0.001	0.001	
Bacillus megaterium NCTC 9848	Aerobic	<0.001	0.66	
Lactobacillus casei	Aerobic	<0.001	0.001	
Staphylococcus albus	Aerobic	<0.001	1.4	Bishop, Pandya,
Staphylococcus albus	Anaerobic	<0.001	0.01	and King,
Sarcina lutea	Aerobic	<0.001	1.8	1962
Clostridium sporogenes	Anaerobic	<0.001	0.001	
Corynebacterium diphtheriae PW8	Aerobic (low Fe content)	<0.001	1.3	
Corynebacterium diphtheriae PW8	Aerobic (high Fe content)	<0.001	3.9	
Mycobacterium tuberculosis	—	<0.0002	—	Lester and
Bacillus mesentericus	—	<0.001	—	Crane, 1959
Corynebacterium diphtheriae, toxic strain	—	—	6.0	Scholes and King, 1963
Micrococcus lysodeikticus	—	—	1.27	Bishop and King, 1962
Gram-negative				
Hemophilus parainfluenzae	—	—	~1.00	White, 1965a
Azotobacter chroococcum	Aerobic	0.41	0.05	
Escherichia coli	Aerobic	0.36	0.32	
Escherichia coli	Anaerobic	0.37	0.28	
Proteus vulgaris	Aerobic	0.65	0.62	
Proteus vulgaris	Anaerobic	0.55	0.51	Bishop et al.,
Pseudomonas pyocyanea	Aerobic	1.45	<0.03	1962; Pandya,
Pseudomonas pyocyanea	Anaerobic	1.35	<0.03	Bishop, and
Aerobacter aerogenes	Aerobic	0.61	<0.05	King, 1961
Pasteurella pseudotuberculosis	Aerobic	0.12	<0.005	
Chromobacter prodigiosum	Aerobic	0.32	<0.010	
Neisseria catarrhalis	Aerobic	1.97	<0.05	
Achromobacter hartlebii	Aerobic	0.42	<0.08	
Azotobacter vinelandii	—	2.6	—	
Pseudomonas fluorescens	—	0.94	—	Lester and
Rhodospirillum rubrum	—	4.3	—	Crane, 1959
Chromatium sp.	—	2.9	—	
Chromatium vinosa	—	2.5	3.5	Osnitskaya et al., 1964
Coryebacterium diphtheriae CN2000	—	—	6.6	Scholes and King, 1965b
Azotobacter vinelandii	—	3.06	—	Jones and Redfearn, 1966
Streptococcus faecalis	—	—	0.13–0.21	Baum and Dolin, 1965
Proteus P_{18}	—	0.31	0.29	Rebel and
Proteus P_{18}, L-form	—	0.63	0.63	Mandel, 1965

The role of the latter is played by organic acids (fumarate) or nitrate. The electron donors are hydrogen and formate. These intestinal anaerobes include *Vibrio succinogenes,* many species of *Bacteroides,* and *Fusiformis nigrescens* (Wolin *et al.,* 1961; White *et al.,* 1962; Lev, 1959). In our view, the discovery of vitamin K in these bacteria (Gibbons and Engle, 1964; Gibbons and Macdonald, 1960) is no more surprising than the presence of cytochromes in these obligate anaerobes, since all these components may be active in a shortened respiratory chain. However, the amount of naphthoquinone in them is so small that it can be detected only by an exceptionally sensitive microbiological method. A real exception is the presence of naphthoquinones in cytochromeless organisms, such as the lactic acid bacterium *S. faecalis* (Baum and Dolin, 1963). The naphthoquinone of this species has been identified as 2-solanesyl-1,4-naphthoquinone, which has 45 carbon atoms in the polyisoprenoid side chain (Baum and Dolin, 1965). Dolin (1963*b*) thinks that quinones may be physiological intermediates in the metabolism of lactic acid bacteria.

The occurrence and role of ubiquinones and naphthoquinones in obligate anaerobes (photosynthetic bacteria) have been discussed in several studies (Osnitskaya *et al.,* 1964; Carr, 1964; Carr and Exell, 1965; Green and Mascarenhas, 1964; Sugimura and Okabe, 1962).

The amount of ubiquinones in bacterial cells varies from 0.30 to 2 μmole per gram of dry weight. The naphthoquinone content lies within practically the same range. These figures are comparable with the ubiquinone content of mitochondria from animal tissues: 2–8 μmole per gram of dry weight (Pumphrey and Redfearn, 1960). The richest producers of vitamin K_2 are *M. phlei* and *B. cereus* strain 35, and the richest producer of ubiquinone (UQ_{10}) is *Pseudomonas denitrificans* (Jacobsen and Dam, 1960; Page *et al.,* 1960).

There have been very few investigations of ubiquinone biosynthesis in bacteria (Burton and Glover, 1965; Rudney and Sugimura, 1961; Azerad *et al.,* 1965; Raman *et al.,* 1965). Brodie's initial opinion (1961*a*) that the synthesis of uniquinones and naphthoquinones, like cytochromes, is poorer in anaerobic conditions has not been confirmed. As Table 6 and the data given below show, anaerobic conditions of cultivation do not significantly reduce the quinone content of most cells. The effect of oxygen pressure during cultivation on the vitamin K_2 content of particles of *B. subtilis* (according to Downey 1964) is illustrated by the following data:

Oxygen pressure during growth of cells, mm Hg	$K_{2(35)}$, μmole/g dry weight
160	0.81
80	0.76
40	0.60
0	0.64

A concentration of ubiquinones or naphthoquinones is found in membranes or membrane fragments containing the respiratory chain. For instance, while *A. vinelandii* cells contain 2.6 μmole of UQ_8 per gram of dry weight, a preparation of the electron-transport particles contains 6.0 μmole of ubiquinone (Lester and Crane, 1959). In *C. diphtheriae* cells, 53% of the vitamin K_2 content is localized in the particles responsible for electron transport (Scholes and King, 1963).

The supernatant fluid of *M. phlei* after fractionation of the electron-transport particles contains only one-tenth of the amount of naphthoquinone which is in the particles. The supernatant fluid of *E. coli* after similar fractionation contains no quinones (Brodie, 1961*b*). Similar information indicating the localization of naphthoquinones in the particles or membranes responsible for electron transport has been obtained by fractionation of the respiratory systems of several bacteria, including *E. coli, M. lysodeikticus, P. pyocyanea, B. subtilis, Proteus* P18, and some nonsulfur purple bacteria (Bishop *et al.,* 1962; Pandya and King, 1962; Carr and Exell, 1965; Rebel *et al.,* 1964). Yet it has been shown for acetic acid bacteria that 40% of the ubiquinones is contained in the membranous fraction and the other 60% in the cytoplasm (De Ley, 1964). That ubiquinone and menaquinone reductases are also localized in the particulate fraction has been shown for *E. coli, B. megaterium, M. lysodeikticus* (Pandya and King, 1966), and *Hydrogenomonas eutropha* (Repaske and Lizotte, 1965).

Despite very intensive research, the functions of ubiquinones and naphthoquinones cannot be regarded as completely clarified. A modern scheme of the mitochondrial respiratory chain ascribes to ubiquinone a role between flavoprotein dehydrogenases and cytochrome *b* (Lehninger, 1964). This opinion is based on the investigations of Hatefi (1963) and Green and Oda (1961). Some authors, however, believe that in intact mitochondria a large part of the endogenous ubiquinone is a side pathway of the electron transport chain (Chance and Redfearn, 1961; Redfearn and Burgos, 1966).

The question of the role of vitamin K in the mitochondrial respiratory chain is a subject of lively discussion. According to Ernster, vitamin K is involved in mitochondrial NADH oxidase in accordance with the initial hypothesis of Martius and Nitz-Litzow (Ernster and Chuan-pu Lee, 1964; Martius and Nitz-Litzow, 1953, 1954; Cunningham et al., 1964). It is still not clear, however, whether vitamin K is involved in the main respiratory pathway or in some shunt mechanism. There is evidence that vitamin K can obtain electrons from vitamin K reductase, which is dissolved in the cytoplasm and is not a component of the mitochondrial electron-transport chain, as had been assumed (Conover and Ernster, 1963; Ernster, 1961; Danielson and Ernster, 1962).

The role of ubiquinones and naphthoquinones in bacterial respiratory chains is being intensively studied. The role of naphthoquinones in the phosphorylating respiratory chain has been thoroughly investigated in the work of Brodie et al. These data are discussed in the next chapter in connection with the question of oxidative phosphorylation.

There have been a few investigations of the role of quinones in substrate oxidation in bacteria. A common feature of these studies is the method employed: Extraction with solvents or irradiation with light (wavelength 360 mμ) will lead to defective respiratory activity. By subsequent addition of ubiquinones or naphthoquinones the respiration (or phosphorylation) is restored. The agent responsible for the restoration is usually regarded as a possible participant in the

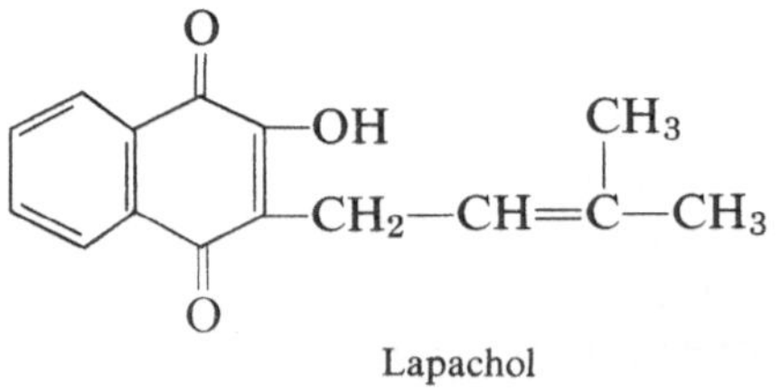

Lapachol

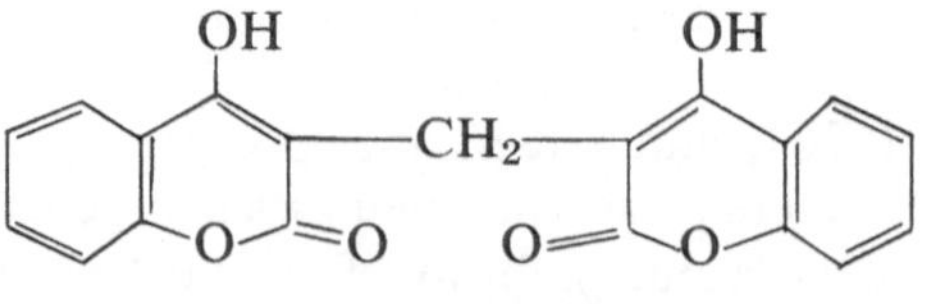

Dicumarol

Table 7. Participation of Ubiquinones and Naphthoquinones in Electron-Transport Chains of Bacteria

Organism (particles)	Restored process	Quinone	Author
Azotobacter vinelandii	Succinate:cytochrome *c* reductase	UQ_8	Temperli and Wilson, 1962
Corynebacterium diphtheriae	Succinic oxidase	$K_{2(45)}$	Scholes and King, 1963
Bacillus stearothermophilus	Succinic oxidase	K_1 or UQ	Downey *et al.*, 1962; Downey, 1963
Bacillus subtilis	NADH/oxidase	$K_{2(35)}$	Downey, 1964
Escherichia coli	Formate:cytochrome *b* reductase	UQ_8 or K_3	Itagaki, 1964
Escherichia coli	Succinic oxidase	UQ_8 or K_2	Kashket and Brodie, 1963*a*
Mycobacterium tuberculosis $H_{37}Ra$	NADH/oxidase	$K_{2(45)}$	Segel and Goldman, 1963
Mycobacterium tuberculosis	NADH/oxidase	$K_{2(45)}$	Kusunose and Goldman, 1963
Mycobacterium phlei	NADH/oxidase	$K_{2(45)}$	Weber, 1963
Mycobacterium phlei	Malate oxidase NADH/oxidase	K_9N	Asano and Brodie, 1956*b*
Hemophilus parainfluenzae	Succinate:NADH:lactate:formate:neotetrazolium reductase	Demethyl K_2	White, 1965*a*
Acetobacter xylinum	Malate oxidase	K_3	Benziman and Perez, 1965
Micrococcus lysodeikticus	NADH/oxidase	K_1 or K_2	Fujita *et al.*, 1966

particular respiratory chain, espcially if the activating quinone is similar to the natural one.

The second method is the use of inhibitors, such as Dicumarol and lapachol, which in their chemical nature are analogs of natural ubiquinones and naphthoquinones.

There is no doubt that this information is of definite value for an understanding of the role of quinones in respiratory chains (Table 7). It has been suggested, however, that the restoration of oxidation by the addition of some quinone is not always of biological significance (Wagner and Folkers, 1963; Redfearn and Pumphrey, 1958; Redfearn *et al.*, 1960). Hence, the information presented below must be treated with some caution.

Vitamin K_1 and UQ_{10} greatly stimulate the respiratory chain in particles (56,000 g) from *B. stearothermophilus* (Downey *et al.*, 1962).

It was later shown that only vitamin K_1 is a definite participant in the respiratory chain (Downey, 1963). Another spore-forming bacillus (*B. subtilis*) was found to contain vitamin $K_{2(35)}$ firmly bound in the particulate fraction (60,000 g and 140,000 g). Vitamin $K_{2(35)}$ is the connecting link between the weakly bound NADH dehydrogenase and the cytochrome chain (Downey, 1964).

Vitamin K_3 has been shown to be a component of the nitrate reductase of *E. coli* (Heredia and Medina, 1960). It is thus of interest that in the same organism the interaction of cytochrome b_1 with formic dehydrogenase is also mediated by vitamin K_3 (Itagaki *et al.*, 1961). Itagaki has shown that neither phospholipids nor neutral lipids extractable with cold acetone are of any significance for the activity of the respiratory chain of *E. coli*. The activity lost on extraction can be restored by vitamins K_3, K_2, or UQ_8 (Itakagi, 1964). Yet Kashket and Brodie (1963*a*) stress that only vitamin K_2 is a definite component of the respiratory chain of *E. coli*. A purified menadione reductase (NADPH:2-methyl-1,4-naphthoquinone oxidoreductase) has been obtained from *E. coli* cells. However, solubilization of this particle-bound enzyme led to the loss of ability to react with natural compounds—UQ_8 and vitamin K_2 (Bragg, 1965*a*). In addition, the menadione reductase from *E. coli* can oxidize NADH with the reduction of a purified preparation of cytochrome b_1 without the addition of quinones. The author regards this as an alternative pathway of NADH oxidation (Bragg, 1965*b*).

Reconstruction of the respiratory chain after extraction of the lipids from *M. phlei* particles with isooctane was effected by the addition of vitamin K_1 (Weber *et al.*, 1958). It is of interest that vitamin K_1 is implicated in the oxidation of malic, fumaric, β-hydroxybutyric, and pyruvic acids, i.e., wherever NADH is generated and oxidized, and is not necessary for the oxidation of succinate (Brodie and Ballantine, 1960). Oxidation is inhibited by competitive inhibitors such as lapachol and Dicumarol (Brodie, 1961*b*). It was later found that particles from *M. phlei* contain vitamin $K_{2(45)}$, which is contained in the NADH oxidase inhibited by irradiation at 360 mμ (Weber *et al.*, 1963).

Experiments involving the addition of several naphthoquinone analogs to a fraction of irradiated cell particles showed that many unnatural quinones can restore oxidation, whereas phosphorylation occurs only when vitamin K_1 is added.

The NADH oxidase of granules from *M. tuberculosis* is inactivated when the lipids are extracted with isooctane and is reactivated by vitamin $K_{2(45)}$ (Kusunose and Goldman 1963). It has recently been shown that NADH oxidase from *M. tuberculosis* actually contains vitamin $K_{2(45)}$ (Segal and Goldman, 1963). Vitamin K_2 is involved in electron transport in particles from *C. diphtheriae* (Scholes and King, 1963, 1965*a*). Investigations of an initial and a mutant strain of *C. diphtheriae* showed that the defective oxidation in the mutant was due to defective synthesis of naphthoquinone (Krogstad and Howland, 1966).

Convincing evidence in support of the participation of demethyl vitamin K_2 in the respiratory chain of *H. parainfluenzae* is given by White (1965*a*). The rate of oxidation and reduction of quinone is correlated with the rate of oxidation and reduction of other components of the respiratory chain. In the case of cyanide inhibition the quinone located between the dehydrogenases and cytochromes *b* and c_1 takes the whole electron flow on itself. Extraction with aqueous acetone inhibits the action of the respiratory chain, which can be regenerated by the addition of demethyl vitamin K_2. The quinone is firmly bound in the membrane and interacts only with membrane-bound dehydrogenases. Detachment of the dehydrogenases by ultrasonic treatment of the membranes leads to inhibition of the respiratory chain (White, 1965*b*).

In the bacterial respiratory chain, vitamin K can be reduced not only by the action of several dehydrogenase components of the chain, but also by specific enzymes—malate:vitamin K reductase and NADH:vitamin K reductase. These enzymes are weakly bound with the respiratory chain and easily pass into the soluble state (Asano and Brodie, 1963, 1964, 1965; Adelson *et al.*, 1964; Benziman and Perez, 1965). The EPR signal found in *M. phlei* particles in the presence of NADH indicates that naphthoquinones are involved in electron transport. The signal owes its origin to enzymatic reduction of naphthoquinones and to the formation of seminaphthoquinones (Weber *et al.*, 1965; Hollocher and Weber, 1965).

It has been established that UQ_8 plays a role in the function of succinate:cytochrome *c* reductase in *A. vinelandii* particles (Temperli and Wilson, 1962).

Summing up, we can say that the establishment of the exact localization and mode of action of ubiquinones and naphthoquinones in bacterial respiratory chains is a problem for future investigation.

CYTOCHROMES

Cytochromes* are the most characteristic components of the respiratory chain. They owe their role to their ability to undergo reversible oxidation and reduction, during which there is a change in valence of the iron contained in the cytochrome molecule (Theorell, 1947).

The mammalian and yeast cytochromes localized in the mitochondrial membranes have been the most studied. The only exception is cytochrome b_5, which is found in the membranes of the endoplasmic reticulum. The composition of the cytochromes in the mitochondria of various animal cells is similar in general features, although there are differences in the properties of individual cytochromes. All the cytochromes found in mitochondria are contained in the respiratory chain and, thus, are functionally connected with respiration and oxidative phosphorylation. This also applies to bacterial cytochromes, but in several groups of bacteria the cytochromes perform additional functions. This is a reflection of the diverse kinds of metabolism in bacteria and the relative simplicity with which the cell can change from aerobic to anaerobic conditions of existence and from autotrophy to heterotrophy.

The earlier view that the presence of cytochromes in bacteria is always connected with electron transport only to oxygen must now be altered. The fact is that the cytochromes bring about electron transport in anaerobic conditions in many species of bacteria, including heterotrophs, chemoautotrophs, and photoautotrophs. In this case nitrates, sulfates, and other compounds can serve as the terminal electron acceptor.

Cytochromes are of common occurrence in obligate anaerobic photoautotrophic bacteria. Here they are implicated in photosynthesis, examination of which is beyond the scope of this work. But among photosynthetic bacteria there are the nonsulfur purple bacteria, which can change over to heterotrophic nutrition in the absence of light and develop a respiratory chain with the reduction of oxygen.

Bacteria from the rumen of mammals comprise a special group. They have become adapted to anaerobic conditions of existence but have not lost their cytochromes. Cells of anaerobic bacteria from the rumen of mammals (*V. succinogenes*) contain cytochromes of

---

* To designate the cytochromes we use the EC enzyme classification and nomenclature proposed by the Enzyme Commission of the International Union of Biochemistry.

groups B and C (Jacobs and Wolin, 1963). Cytochromes are absent only in obligate anaerobic fermentation bacteria—butyric acid bacteria, and in facultative anaerobes—lactic acid bacteria (streptococci, pneumococci).

The cytochrome content is fairly high in a number of the obligate aerobes investigated. For instance, *Aerobacter aerogenes* and *Azotobacter chroococcum* cells contain ten times more cytochromes per unit dry weight than yeast cells (L. Smith, 1954*b*). Nevertheless, there is no direct relationship between the respiration rate of bacteria and the cytochrome content, although the turnover number of several bacterial cytochromes even exceeds the number in yeast and heart-muscle mitochondria (L. Smith, 1954*a, b*). Experiments with inhibitors and kinetic investigations have shown that in most aerobic bacteria which contain a cytochrome system, electron transport from substrate to oxygen is effected through this system.

Two features of bacterial cytochrome systems can be noted: the variation of the qualitative and quantitative composition of the cytochromes and the presence of several terminal oxidases (Smith, 1963; White and Smith, 1962, 1964). In addition to the typical mitochondrial set of components (a, a_3, b, c_1, c), as in *B. subtilis* and *S. lutea,* for instance, several bacteria contain these cytochromes in various combinations with cytochromes a_1, a_2, o, b_1, b_4, c_3, c_4, and c_5 (Chin, 1952; Smith, 1963).

It is difficult to investigate the cytochromes in bacterial cells and cell fractions by conventional spectrophotometric technique because of the great turbidity of these preparations and the low pigment content. However, the development of highly sensitive differential spectrophotometers (Chance, 1951, 1952) has permitted the successful investigation of bacterial cytochromes in suspensions of whole cells or homogenates. The technique of differential spectrophotometry has established the presence of cytochromes and their relative amounts in numerous species of bacteria (L. Smith, 1954*a, b, c*). A differential spectrophotometer similar to Chance's instrument has been designed by Borisov and Mokhova and has been used to study the cytochrome content of microorganisms (Borisov and Mokhova, 1964; Mokhova, 1965; Lisenkova and Mokhova, 1964; Pomoshchnikova *et al.,* 1966; Ermachenko *et al.,* 1966; Lisenkova and Lozinov, 1966).

Detection of cytochromes is based on an analysis of their absorption spectra (the α- and β-bands at 550–650 mμ, and the γ-band at 400–500 mμ). Identification of cytochromes involves investigation of their absorption maxima in the oxidized and reduced forms and the absorption maxima of compounds of their hemes with cyanide and pyridine, pyridine ferrohemochromes and cyanide ferrochemochromes. Cytochromes can be divided into four main groups—*A, B, C,* and *D*—according to the nature of the heme of the prosthetic group (Faulk, 1964; Blyumenfel'd and Purmal', 1964). Special tests are used to determine the group to which a cytochrome belongs (Classification and Nomenclature of Enzymes, 1962).

The terminal oxidases of bacteria are those which interact with the final electron acceptor (O_2, nitrate, sulfate, etc.). In contrast to the mitochondrial cytochrome oxidase $a + a_3$, the terminal oxidases of bacteria are diverse. They include cytochromes a, a_1, a_2, a_3 and o, which are characterized by definite α- and γ-bands (Table 8). As Table 8 shows, type A cytochromes in the reduced state have absorption bands in the red region (590–635 mμ).

Cytochrome o is the terminal oxidase in *A. suboxydans* and *Micrococcus pyogenes* var. albus (*Staphylococcus albus*). The absorption bands of cytochrome o resemble the spectrum of hemoglobin. Its prosthetic group is protohemin (L. Smith, 1954a). Cytochrome o interacts with oxygen and carbon monoxide; the compound with carbon monoxide is degraded by light.

Long before the mitochondrial cytochromes a_3 and a were separated chemically, research on type A bacterial cytochromes showed that they each could exist separately, as in *M. pyogenes* var. albus (L. Smith, 1955, 1963), thus bearing out Keilin's view that cytochromes a and a_3 were not the same compound (Keilin and Hartree, 1939).

Some bacteria have only one oxidase, while others have two or three. The occurrence and combined occurrence of various terminal

Table 8. Position of Maxima in Absorption Spectra of Type A Cytochromes (L. Smith, 1955a,b)

	Pigment in reduced form, mμ	
Cytochrome	Visible region, α-band	Soret region, γ-band
a	605	439*
a_1	590	440
a_2	635	?
a_3	605	445

*According to Horie and Morrison, 1964.

oxidases in bacteria have been reviewed by L. Smith (1963). It has been found that *Acetobacter pasteurianum* and *Nitrobacter* contain cytochrome a_1 as a terminal oxidase (Castor and Chance, 1955; Lees and Simpson, 1957). In other bacteria (*A. peroxydans* and *A. chroococcum*), cytochrome a_1 is found along with cytochrome a_2. Cytochrome a_2 does not occur in the absence of cytochrome a_1. Cytochromes a_1 and a_2 are often found in conjunction with cytochrome o, as in *P. vulgaris, E. coli, A. aerogenes,* and *A. vinelandii*. Cytochromes a, a_2, and a_3 are found together in *Pseudomonas riboflavina* (Arima and Oka, 1956b). Cytochromes a_3, a_2, and a_1 react with CO, CN$^-$, and O_2. It has been found that interaction with CO is more characteristic for the spectroscopic identification of bacterial oxidases, since compounds with cyanide can give catalases, which are often present in bacteria (Castor and Chance, 1959). Moreover, cytochrome a as a terminal oxidase does not interact with cyanide.

In some bacteria (*B. subtilis, S. lutea*) the terminal oxidases consist of cytochromes $a + a_3$. Their properties, however, are not identical with those of mitochondrial cytochrome oxidase, since they do not oxidize cytochrome c from tissues of animal origin, i.e., they do not exhibit cytochrome c oxidase activity. Hence, the absence of such activity in other bacteria where the terminal oxidases are different from $a + a_3$ should cause no surprise (L. Smith, 1954a, b; Yamanaka and Okunuki, 1964).

The conditions leading to predominant biosynthesis of a particular bacterial oxidase have been investigated. Oxygen deficiency in the medium stimulates the formation of cytochrome a_2, as has been shown for *E. coli, A. aerogenes,* and *H. parainfluenzae* (Moss, 1956; White and Smith, 1962). It has been shown for four species of bacteria (*E. coli, P. vulgaris, A. aerogenes,* and *P. riboflavina*) that cytochrome a_2 is absent in the logarithmic phase and appears in the stationary phase. The authors attribute this to the reduction in the oxygen content of the medium in the stationary phase (Castor and Chance, 1959; Arima and Oka, 1965b). The relative amounts of terminal oxidases can vary in one species of bacteria with the cultivation conditions, as White and Smith (1962) showed for the facultative anaerobe *H. parainfluenzae*. Cytochrome o predominates in a well-aerated culture, cytochrome a_2 predominates with limited aeration, and in anaerobic conditions with nitrate present there is an increase in a_1. According to White and Smith (1962), cytochrome a_1 is associated with nitrate reduction. It has been shown that one of the terminal oxidases, the

Table 9. Type C Bacterial Cytochromes

Cyto-chrome	Maxima in reduced state, $m\mu$ (α-, β-, and γ-bands)	Source of cytochrome	Author
c	551, 521, 416	*Pseudomonas aerugi-nosa*	Verhoeven and Takeda, 1956 Rose and Ochoa, 1956
c_1	554, 525, 416	*Pseudomonas aerugi-nosa*	Horio, 1958*a,b;* Horio *et al.,* 1960
c_2	550, 521, 416	*Rhodospirillum ru-brum*	Horio and Kamen, 1961
c_3	553, 525, 419	*Desulfovibrio desul-furicans*	Postgate, 1956 Ishimoto *et al.,* 1954
c_4	551, 522, 416	*Azotobacter vine-landii*	Tissières, 1956
c_5	555, 526, 420		Tissières and Burris, 1956

nitrate reductase of *P. aeruginosa,* appears when the bacteria are grown on nitrate in anaerobic conditions (Pseudomonas cytochrome c_{551}; nitrite:O_2 oxidoreductase). The presence of oxygen induces the biosynthesis of another terminal oxidase, cytochrome oxidase (cytochrome c:O_2 oxidoreductase, EC 1.9.3.1, Azoulay, 1964; Azoulay and Couchoud-Beaumont, 1965). Thus, oxygen and nitrate are competitive electron acceptors in the cells of facultative anaerobes.

The relative amounts of cytochromes c and c_1 in bacteria vary. In *M. tuberculosis,* cytochrome c_1 is found in the absence of cytochrome c (Goldman *et al.,* 1963). Cytochromes c_2, c_3, c_4, and c_5 are found only in bacteria (Table 9).

Cytochromes of group C are present in the respiratory chains of bacteria where the terminal electron acceptor is not oxygen, but other oxidants (sulfate, nitrate, oxidants formed in photosynthesis).

The prosthetic group of bacterial cytochromes C, like that of heart-muscle cytochrome, is mesoheme. However, the protein components of these cytochromes differ considerably, as is indicated by the diversity of oxidation–reduction potentials of bacterial cytochromes C, the amino-acid composition of the protein components, and the enzymatic specificity (see Table 11).

Cytochrome c of *A. suboxydans* is not oxidized by mitochondrial cytochrome oxidase (Iwasaki, 1960). Only some of the type C cytochromes that have been isolated are oxidized by mitochondrial cytochrome oxidase and then only very slowly. Type C bacterial cytochromes are rapidly oxidized and reduced by oxidases and reductases contained in preparations obtained from the same organism (Kamen and Vernon, 1955; Vernon, 1956; Tissières, 1956). Cytochromes c_4 and c_5 of *A. vinelandii* are rapidly oxidized by oxidases of these bacteria, but are not oxidized by oxidases of other bacteria and mammals (Tissières and Burris, 1956). It is believed that none of the bacterial cytochromes of group *C* is identical with heart-muscle cytochrome *c* in physicochemical and biological properties (see Table 11). Recent research has shown a connection between individual cytochromes *C* and their physiological function. Tissières, 1956 showed that cytochromes c_4 and c_5 are involved in the succinic oxidase and NADH oxidase systems of *A. vinelandii*. Cytochrome c_{553} functions in the oxidation of ethanol in *A. peroxydans* (Nakayama, 1961). This cytochrome is also implicated in the oxidation of primary, secondary, and poly alcohols (De Ley and Kersters, 1964; Kersters *et al.*, 1965). Cytochrome *c* is part of the nitrite oxidase system in *Nitrobacter* (Butt and Lees, 1958; Aleem and Nason, 1959; Nason and Aleem, 1961). Cytochrome c_2 is found in photosynthetic bacteria. In Smith's opinion (1957), this cytochrome is not contained in the respiratory chain. This pigment has been isolated from *R. rubrum* (Smith, 1957; Horio and Kamen, 1961). Type *C* cytochromes are implicated in nitrate reduction in *Micrococcus denitrificans, Achromobacter fischeri, P. aeruginosa, E. coli* (Sato, 1956; Sadana and McElroy, 1957; Verhoeven and Takeda, 1956; Wimpenny *et al.*, 1963). The cytochrome *C* content of these organisms is increased when the cells are put into the most favorable conditions for the action of nitrate-reducing systems (Verhoeven and Takeda, 1956). A preparation of particles from *V. succinogenes* has twice as great a cytochrome *c* content if the cells are grown on nitrate (Jacobs and Wolin, 1963). In the anaerobic bacterium *Desulfovibrio desulfuricans*, cytochrome c_3 is involved in the reduction of sulfite, sulfate, and thiosulfate or dithionite (Postgate, 1956; Ishimoto *et al.*, 1957; Peck, 1959).

Bacterial cytochromes of type *B* also vary according to the organism (Table 10). A common feature of bacteria cytochromes of group *B*, and of mitochondrial cytochrome *b*, is their inability to interact with CO, HCN, or HN_3 (Pappenheimer, 1955; Pappenheimer *et al.*, 1962; Jackson and Lawton, 1959). Investigations of the oxidation–reduction potential of several bacterial cytochromes of group *B* have shown that they precede cytochrome *C* in the respira-

Table 10.   Type B* Bacterial Cytochromes

Cyto-chrome	Position of absorption maxima in reduced form, $m\mu$			Microorganism
	α	β	γ	
b	564	523	422	*Bacillus subtilis*
	565	523	427	*Staphylococcus albus*
	562	532	428	*Bacterium anitratum*†
	562	523	430	*Sarcina lutea*
b_1	560	530	430	*Propionibacterium pentosaceum*‡
	560	525	430	*Micrococcus lysodeikticus*§
	560	530	430	*Aerobacter aerogenes*
	560	533	432	*Escherichia coli*
	560	530	428	*Azotobacter chroococcum*
	560	524	429	*Corynebacterium diphtheriae*
	559	528	426	{ *Micrococcus denitrificans*¶ { *Pseudomonas denitrificans*
Type B ?	554	523	428	*Acetobacter pasteurianum*
	554	525	429	*Acetobacter suboxydans*

* Pappenheimer, 1955.
† Hauge, 1960.
‡ Lara, 1959.
§ Jackson and Lawton, 1959.
¶ Vernon, 1956.

tory chain. It has been shown for the bacterial respiratory chain that group B cytochromes are on the pathway of electron transfer from NADH, as well as from succinate and several other respiratory-chain substrates. The implication of cytochrome B in succinate oxidation has been shown for *C. diphtheriae, Bacterium tularense,* and *P. pentosaceum,* and in NADH oxidation for several bacteria, including *M. lysodeikticus, H. parainfluenzae, C. diphtheriae, E. coli,* and *M. phlei* (Pappenheimer *et al.,* 1962; Lara, 1959; Wadkins and Mills, 1956; Ishikawa and Lehninger, 1962; Gel'man *et al.,* 1963; White, 1965a; Itagaki, 1964; Asano and Brodie, 1964).

Vernon (1956) isolated cytochrome b_1 from the nitrate-reducing bacteria *M. denitrificans* and *P. denitrificans.* It was reduced by succinate and NADH and oxidized by nitrate in the presence of cell extracts. In *E. coli* and *Achromobacter,* cytochrome b_1 is also involved in nitrate reduction (Taniguchi *et al.,* 1956; Sato, 1956; Itagaki *et al.,* 1961; Egami *et al.,* 1961; Arima and Oka, 1965a). There are reports that cytochrome b_1 is part of succinic and formic dehydrogenase systems (Linnane and Wrigley, 1963; Asnis et al., 1956). Cytochrome b_4 from a halotolerant strain of Micrococcus is associated with electron transfer from NADH and succinate to nitrite and hydroxylamine (Hori, 1961).

In some bacteria cytochrome B is autoxidizable, although its interaction with oxygen is much weaker than for cytochrome oxidase (Jackson and Lawton, 1959; Pappenheimer et al., 1962; Itagaki et al., 1961). In the toxic strain of C. diphtheriae ($PW8_sPd$), cytochrome b_1 is the only cytochrome and acts as the terminal oxidase (Pappenheimer et al., 1962).

In bacteria, as in mitochondria, cytochromes are localized in the membranous structures. The strength of the connection of the cytochrome components with the bacterial membranes varies considerably. Hence, there are no standard procedures for their solubilization and subsequent purification. The extraction of a particular cytochrome from each organism requires its own special procedure. For instance, Jackson and Lawton (1959) could not extract cytochrome b from M. lysodeikticus by the methods (disruption of cells with aluminum oxide and extraction with neutral phosphate buffer) used successfully by Vernon (1956) to solubilize the cytochromes of M. denitrificans and P. denitrificans. Jackson and Lawton also reported that neither desoxycholate nor Tween could solubilize the cytochrome of M. lysodeikticus. This cytochrome can be solubilized by treating the membranes with the cationic detergent cetyl trimethylammonium bromide. Cytochrome b in H. parainfluenzae is also very firmly bound with the cell membranes: it is not released from the membranes by repeated freezing and thawing, whereas cytochrome c_1 in this organism is partially solubilized by this treatment (White and Smith, 1964). Nevertheless, mainly through the work of Japanese researchers, progress has been made in the purification of many bacterial cytochromes. Some cytochromes have been obtained in crystalline form and the amino-acid composition of some has been determined. These include the cytochrome c and cytochrome oxidase from P. aeruginosa and the cytochrome c from P. fluorescens.

In this book we cite recent work on the solubilization and purification of several cytochromes. The literature up to 1958 has been reviewed by Dolin (1963a), L. Smith (1963), and Newton and Kamen (1963). The characteristics of purified cytochromes of photosynthetic bacteria have also been reviewed (Kamen et al., 1963; Kondrat'eva, 1963). As Table 11 shows, most of the purified bacteria cytochromes

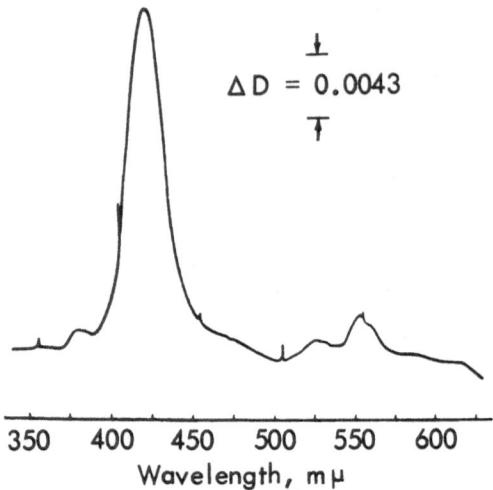

$$\Delta D = 0.0043$$

350   400   450   500   550   600
Wavelength, mμ

Fig. 15. Absorption spectrum of intact *Pseudomonas fluorescens* cells. 500 μg of protein per milliliter (from Lysenkova, 1965).

belong to group *C*. Like the cytochrome *c* from mitochondria, they are probably the cytochromes the least firmly attached to the structure.

After disruption of the cell walls by any method (enzymatic, ultrasonic or mechanical), many bacterial cytochromes of group *C* pass into solution on extraction with buffers or weak solutions of bases or acids. The usual methods of salting out and purifying are then used. However, some cytochromes of group *C* and a number of those of groups *A* and *B,* are not brought into solution by such treatment (Horio, 1958*a*; Vernon and Mangum, 1960). Hence, detergents, bile salts, and lipolytic enzymes are used to solubilize them. To solubilize group *C* cytochromes from *B. megaterium* and *B. subtilis,* Vernon and Mangum (1960) employed digestion of a particulate preparation with lipase for 12 hr. Cytochrome b_1 can be brought into solution from *E. coli* particles by 18-hr treatment with snake venom in the presence of 0.1% sodium desoxycholate (Fujita *et al.,* 1963).

The terminal oxidases of *A. vinelandii, P. denitrificans, M. denitrificans,* and certain other species have been partially purified (Tissières *et al.,* 1957; Tissières, 1956; Vernon, 1956; Kamen and Vernon, 1954*a, b*; Layne and Nason, 1958; Vernon and White, 1957). Among group *A* cytochromes, only the cytochrome oxidase (nitrite

Table 11. Properties of Bacterial Cytochromes

Source of cytochrome	Cytochrome	Absorption maxima in reduced form, mμ	Isoelectric point, pH	Molecular weight	Redox potential E_0' at pH 7, v	Autoxidizability	Author
Pseudomonas aeruginosa	c*₅₅₁†	551, 521, 416, 316, 290, 280	4.7	8,100	+0.286	—	Horio et al., 1960; Horio et al., 1958a, b; Coval et al., 1961
Pseudomonas aeruginosa	Cytochrome oxidase (nitrite reductase) a₂ c†	655, (630) 554, 549, 523, 420	5.8	90,000	—	—	Horio et al., 1958a, b; Horio et al., 1961a; Yamanaka and Okunuki, 1963a, b; Yamanaka, 1964; Yamanaka et al., 1963, 1962; Ambler, 1963
Pseudomonas fluorescens	c*₅₅₁†	551, 520, 416, 316	4.7	9,000	—	—	Deeb and Hager, 1964
Escherichia coli	b₅₅₇†	557, 527, 425	—	62,000	−3.400	Autoxidizable	Fujita and Sato, 1963
Escherichia coli	c₅₅₂	552, 523, 420	—	26,000	−0.200	Autoxidizable	Vernon and Mangum, 1960
Bacillus megaterium	c₅₅₀	550, 520, 415	—	—	+0.250	Autoxidizable	Vernon and Mangum, 1960
Bacillus subtilis	c₅₅₀	550, 520, 415	—	—	—	—	Sutherland, 1963
Bordetella pertussis, B. parapertussis	c₅₅₃	553, 522, 416	—	—	+0.192	Slowly autoxidized	Sutherland, 1963
B. bronchioseptica	c₅₅₀	550, 520, 418	—	—	+0.259	Not autoxidizable	
Desulfovibrio desulfuricans	c₅₅₃*	553, 525, 419	10.5	12,000	−0.205	Autoxidizable	Horio and Kamen, 1961; Postgate, 1956; Ishimoto et al., 1954; Coval et al., 1961: Takahashi et al., 1959
Desulfovibrio gigas	c₅₅₃	553, 525, 419	5.2	—	−0.220	—	Le Gall et al., 1965
Thiobacillus X	c₅₅₀	—	—	—	+0.200	Slowly autoxidized	Trudinger, 1958, 1961, 1964
	c₅₅₃	553, 522, 403	—	—	+0.210	Rapidly autoxidized	
	c₅₅₇	—	—	—	+0.155	Very rapidly autoxidized	
Thiobacillus ferrooxidans	c₅₅₂	552, 523, 417	—	—	+0.310	—	Vernon et al., 1960
Thiobacillus denitrificans	c₅₅₂	552, 522, 415, 316	≈10.2	—	+0.270	Not autoxidizable	Aubert et al., 1959
Nitrobacter agilis	c₅₅₀	550, 521, 416	6.5	—	+0.250	Not autoxidizable	Butt and Lees, 1958

*The amino-acid composition of the protein component of the cytochrome has not been determined.
† Obtained in crystalline form.

reductase) of *P. aeruginosa* has been purified sufficiently and crystal-lized. The enzyme contains two hemes (c and a_2) but does not contain copper, although its biosynthesis is stimulated by the addition of copper sulfate to the cultivation medium. Heme a_2 is a chlorine-dihy-droporphyrin derivative (Yamanaka, 1964; Barret, 1956), whereas the prosthetic group of cytochromes a, a_1 and a_3 in heme a (ferro-formylporphyrin) (Falk, 1964; Lemberg *et al.,* 1961; Morrison and Stotz, 1961). Purification of the nitrite reductase of *A. fischeri* showed that this enzyme differed from the nitrite reductase of *P. aeruginosa.* It contained only one heme (C-type) and in the reduced form had absorption maxima at 420, 522 and 552 m (Prakash *et al.,* 1966). An-other terminal oxidase of *P. aeruginosa* (cytochrome $c:O_2$ oxidore-ductase, EC 1.9.3.1) (Azoulay and Couchoud-Beaumont, 1965) has been partially purified. The cytochrome spectrum had maxima at 420, 522, and 595 mμ. The enzyme contained type A heme.

Cytochrome b_1 of *E. coli* has been isolated, purified, and charac-terized (Deeb and Hager, 1964). It has been obtained in crystalline form and its molecular weight determined as 62,000. The oxidation–reduction potential of the crystalline cytochrome is -0.340 v, where-as the unpurified preparation has a potential of -0.010 v. In extract-ing this cytochrome, Fujita *et al.* (1963) noted that although treat-ment of the particles with snake venom and subsequent extraction with cholate resulted in isolation of cytochrome b_1, the activity of pyruvate oxidase was destroyed. Hence, Deeb and Hager used soni-cation to extract this cytochrome from the particles.

Hori (1961) managed to partially purify two b_4-type cytochromes from a halotolerant species of micrococcus. It was later found, how-ever, that these preparations contained an admixture of cytochrome c (Falk *et al.,* 1961). Pappenheimer (1955) partially purified cyto-chrome b_1 from *C. diphtheriae.* Jackson and Lawton (1959) partially purified cytochrome b from *M. lysodeikticus.* Purified bacterial group B cytochromes, like the mitochondrial pigment, contain protoheme IX as a prosthetic group (Vernon, 1956; Deeb and Hager, 1964; Jacobs and Wolin, 1963).

Solubilized and purified cytochromes largely have been those of type C. Cytochromes c_{551} from *P. aeruginosa* and *P. fluorescens* have been thoroughly investigated. They have been obtained in crystalline form and their amino acid composition has been de-termined (Ambler, 1963; Coval *et al.,* 1961). Ambler crystallized cyto-chrome c_{551} from *P. fluorescens* and showed that its protein part consists of a single peptide chain with 82 amino acid residues and has a molecular weight of 9,000. There is one heme group per molecule. This cytochrome has properties very similar to those of cytochrome c_{551} from *P. aeruginosa* (Ambler, 1963; Horio *et al.*, 1960). The isoelectric points of these cytochromes lie in the acid region (pH 4.7).

The prosthetic groups of type C bacterial cytochromes have been analyzed for several purified preparations from species of *Azotobacter, Pseudomonas,* and *Micrococcus,* as well as from *E. coli* (L. Smith, 1963; Fujita *et al.*, 1963). These groups proved to be identical with those of the well-investigated mitochondrial cytochrome *c.*

The prosthetic group of group C cytochromes is a substituted mesoheme. Cytochromes c_4 and c_5 from *A. vinelandii* possess a pyridine ferrohemochrome similar to that of mammalian cytochrome *c* (Tissières, 1956). The pyridine ferrohemochrome and cyanide ferrohemochrome *c* from iron bacteria, *B. subtilis* and *B. megaterium* are similar to those compounds from mammalian mitochondria (Vernon *et al.*, 1960; Vernon and Mangum, 1960). An analysis of the protein components of these cytochromes, however, showed that they were different from those of the mitochondrial cytochromes. The isoelectric points, electrochemical potential, electrophoretic mobility, and enzymatic specificity were different.

As distinct from mitochondrial cytochrome *c*, with isoelectric point at pH 10, group C bacterial cytochromes often have isoelectric points at acid pH (see Table 11). The oxidation–reduction potentials of group C cytochromes vary in a wide range, depending on the species of bacteria. This indicates the diversity of the protein components of the cytochromes (see Table 11).

Vernon *et al.* (1960) extracted cytochrome *c* from the iron bacterium *Thiobacillus ferrooxidans.* He found that the potential of cytochrome *c* is $+0.310$ v, whereas the potential of iron is $+0.770$ v. It is still unclear how bivalent iron can reduce cytochrome *c*, the potential of which is lower. It has been shown, however, that the potential of this cytochrome is increased at low pH, which are physiological for this species.

Cytochrome c_{552} from *Thiobacillus denitrificans* is similar in many physicochemical properties to cytochrome *c* from mammalian mitochondria. Its isoelectric point is at pH 10.2 (Aubert *et al.*, 1959). Cytochrome c_3 from *D. desulfuricans* has absorption maxima at 553, 525, and 415 mμ. Its molecular weight is 12,000; one cytochrome molecule contains two heme groups. This cytochrome has a very low

Table 12. Dehydrogenase—Cytochrome Complexes of Bacteria

Organism	Complex	Author
Corynebacterium diphtheriae	Succinic dehydrogenase and cytochrome *b*	Pappenheimer and Hendee, 1949 Pappenheimer *et al.*, 1962
Propionibacterium pentosaceum	Succinic dehydrogenase and cytochrome *b*	Lara, 1959
Escherichia coli	Formic dehydrogenase and cytochrome b_1	Linnane and Wrigley, 1963
Escherichia coli	Formic dehydrogenase, nitrate reductase, and cytochrome b_1	Itagaki, Fujita, and Sato, 1961; Fujita *et al.*, 1963
Rhodospirillum rubrum	Succinic dehydrogenase and cytochrome *b*	Taniguchi and Kamen, 1965
Bacterium anitratum	Glucose dehydrogenase and cytochrome *b*	Hauge, 1960
Acetobacter sp.	Alcohol dehydrogenase and cytochrome c_{553}	Nakayama, 1961; Takeyoshi, 1961
Acetobacter suboxydans	Lactic dehydrogenase and cytochrome *c*	Iwasaki, 1960

potential (-0.205 v) and is auto-oxidizable. Analysis of the amino-acid composition showed that cytochrome c_3 contains no tyrosine. This accounts for the absence of absorption at 280 mμ (Takahashi *et al.*, 1959).

When respiratory chains are fractionated bacterial dehydrogenases are often found firmly bound with cytochrome components (Table 12). Many authors have observed a very close structural and functional connection between bacterial cytochrome *B* and succinic dehydrogenase (Lara, 1959; Pappenheimer *et al.*, 1962; Fellman and Mills, 1960). A similar connection has been observed for mitochondrial succinic dehydrogenase from heart muscle (Ziegler and Doeg, 1959). Isolation of complexes of dehydrogenases and cytochromes could lead to a better understanding of the molecular architecture of membranes, since there is reason to think that such complexes exist in the cell *in vivo*. The strong bond of dehydrogenases with group *B* cytochromes may indicate that in the cytochrome series this component is contiguous with the dehydrogenase part of the respiratory chain. This is consistent with the low oxidation-reduction potential of this cytochrome. In some cases dehydrogenases form complexes with type *C* cytochrome (see Table 12). There are data which indicate that group *B* cytochromes are absent in the respiratory chains of these bacteria, and in this case cytochrome c is often associated purely spatially with the dehydrogenase (Nakayama, 1961; Takeyoshi, 1961; Iwasaki, 1960). The interaction of dehydrogenases with the cytochrome components of the respiratory chain occurs only as long as there is a spatial connection between the dehydrogenases and cytochromes on the bacterial membranes (White, 1964; Smith, 1962). Any treatment which destroys this connection and converts the dehydrogenase to the soluble state (ultrasound, lysis, or osmotic shock) prevents the oxidation of substrates (Gel'man *et al.*, 1963; Oparin *et al.*, 1960).

Sometimes the complexes of dehydrogenases and cytochromes contain lipids. The lipid content of the complex of formic dehydrogenase and cytochrome b_1 from *E. coli* is 3% (Linnane and Wrigley, 1963). Lipids are found in a complex with cytochromes in electron transport particles from *B. cereus* (Doi and Halvorson, 1961). Purification of several bacterial cytochromes has also revealed their connection with lipid material (Kamen *et al.*, 1963). It has been shown that the lipid fraction extracted from cells is essential for the reduction of cytochrome b_1 under the action of formate dehydrogenase from *E. coli* (Itagaki *et al.*, 1961).

In cases where bacterial membranes contain yellow pigments there is often a strong complex of carotenoids and cytochromes. A similar complex has been investigated in the membranous fraction of *S. lutea* (Mathews and Sistrom, 1959). Jackson and Lawton (1958, 1959) found a strong complex between cytochrome b and a carotenoid in *M. lysodeikticus*.

A thorough investigation of the attachment of cytochromes to the bacterial lipoprotein membranes should include study of their connections with lipids, carotenoids, and adjacent components of the electron-transport chain.

An important feature of bacterial cytochromes is their great variation with the composition of the medium, aeration, and phase of development of the culture. The change in synthesis of terminal oxidases in relation to cultivation and aeration conditions was mentioned above.

In a series of investigations of *H. parainfluenzae*, White and Smith (1964) studied the change in the proportions of respiratory chain components due to cultivation conditions. They observed a 10- to 30-fold variation in the concentration of various cytochromes, but the respiratory activity in this case was very slightly altered, and the limiting factor was the retention of the link of the dehydrogenases with the cytochrome part of the respiratory chain rather than the cytochrome content.

The connection between the partial pressure of oxygen in the culture medium, cytochrome content, and respiratory activity has been the subject of many investigations. No relationships common to all species have been found. In *B. subtilis*, these three factors are directly related (Table 13). Similar observations have been made for *Staphylococcus epidermidis* cells. Under anaerobic cultivation conditions, only traces of cytochromes can be found and respiration on glucose is reduced by a factor of 10 (Jacobs and Conti, 1965). The authors believe that the presence of oxygen is essential for the biosynthesis of heme. This opinion is confirmed by their data showing that the cytochrome content and respiratory activity of bacteria grown anaerobically in a hemin-containing medium are not reduced. When *B. cereus* is grown in anaerobic conditions, synthesis of cytochromes *a* and *c* is reduced. This cytochrome deficiency, however, is not associated with any reduction of respiratory activity, as in the case of *B. subtilis* and *S. epidermidis* (Schaeffer, 1952).

In many cases there is a reduction of cytochrome content when the medium is deficient in iron. This has been shown for *C. diphtheriae, Aerobacter indolegenes,* and *A. aerogenes* (Mitsuhashi *et al.,* 1956; Tissières, 1951; Waring and Werkman, 1944).

In the analysis of bacterial cytochromes, it is important to take the age of the culture into account. The cytochrome content usually increases after the period of maximal growth (Gibson, 1961). In *A. aerogenes,* the cytochrome content after 48-hr growth is twice that after 20 hr (Smith, 1954c). The oxidase and cytochrome *c* content of *H. parainfluenzae* is higher during the stationary phase (White, 1963).

Table 13.   Effect of Oxygen Pressure on Oxidative Compounds of Bacillus subtilis (according to Downey, 1964)

Oxygen pressure during growth, mm Hg	Cytochromes per 100 mg of protein		Oxygen absorption, μg O_2 per mg of protein per hr
	c, $E_{550-574}$	a_3, $E_{600-610}$	
160	20.0	1.7	135.0
80	14.6	1.1	46.3
40	3.1	0.1	2.2
0	0.1	0.1	0

Different strains of the same bacterium may have a different set of cytochrome components. In an investigation of *C. diphtheriae* Pappenheimer *et al.* (1962) found that the respiratory chain of the fast-growing strain $PW8_sP$ contained a complete set of cytochromes: b_{564}, c_{552}, a, a_3, and a factor between b and c. This factor was absent in a slow-growing strain. The toxic strain possessed a single cytochrome (cytochrome b), which acted as a terminal oxidase. Some differences between the cytochrome sets of an initial strain and a mutant of *H. parainfluenzae* have been observed (White and Smith, 1964).

The change in respiratory pathways when conditions change is one example of the great flexibility of bacterial cells (Kluyver and Van Niel, 1959). When cultivated in anaerobic condition with nitrate in the medium, many obligate aerobic bacteria change over from oxygen to "nitrate respiration," with a slightly different set of respiratory chain components. Such changes have been observed several times in many species of *Pseudomonas*, in *E. coli*, *A. vinelandii* and other bacteria (De Ley, 1964; Yamanaka, 1964; Taniguchi and Ohmachi, 1960).

From the evolutionary viewpoint, nitrate respiration in bacteria is believed to be earlier than oxygen respiration (Yamanaka, 1964). This feature of bacteria is evidence of their evolutionary antiquity. For example, one thousand investigated species of the family Enterobacteriaceae can reduce nitrates (Daubner, 1962).

Cells and spores (or cysts) often differ in their respiratory pathways. Doi and Halvorson (1961) conducted a comparative investigation of the respiratory enzymes in the vegetative cells and spores of *B. megaterium*. The cells contained a typical respiratory chain, and the spores contained a flavin oxidase. The spores lacked cytochromes and ubiquinones (King *et al.*, 1961). In *Myxococcus xanthus*, Dworkin and Niederpruem (1964) found a similar alternation of respiratory pathways—cytochrome in cells, flavin in microcysts.

Thus, in work with bacterial cytochromes, cultivation conditions must be carefully controlled if reproducible results are to be obtained. The preceding account illustrates the great lability of bacterial respiratory chains as compared with mammalian mitochondria, where there is a more rigid pattern of enzymatic reactions and a correspondingly less variable set of respiratory components (Green, 1964*a*, *b*; Green and Fleischer, 1964).

Since photoautotrophic and chemoautotrophic bacteria form a

group characterized by special kinds of energy metabolism, it is expedient to discuss their cytochrome components. Cytochromes are found in the chromatophores of photosynthetic bacteria, but their role in photosynthesis and photosynthetic phosphorylation is not part of our problem. Comprehensive reviews on this topic have been published (Kondrat'eva, 1963; Sisakyan and Bekina, 1964; Newton and Kamen, 1963; *Bacterial Photosynthesis,* 1963).

We are interested only in the facultative, photoheterotrophic, nonsulfur purple bacteria of the family Athiorhodaceae. In this family only some species of *Rhodospirillum* and *Rhodopseudomonas* are facultative aerobes that develop in darkness, obtaining energy by oxidation of organic acids, amino acids, sugars, acetone, molecular hydrogen, or certain inorganic sulfur compounds (Kondrat'eva, 1963; Clayton, 1955; Morita and Conti, 1963). The composition of their respiratory chains is of interest. For a long time they were believed to have a "shortened" chain of carriers in which type *A* cytochromes were absent, so that these bacteria were insensitive to cyanide and carbon monoxide. This is probably true of obligate anaerobic photoautotrophs. The presence of type *C* cytochrome in photosynthetic bacteria is essential. The first bacterial cytochrome *c* was isolated from *R. rubrum* (Kamen and Vernon, 1954*a, b*). *R. rubrum* and *R. spheroides* contain type *B* cytochromes (Kamen, 1962; Newton and Kamen, 1963). In addition, *R. rubrum* contains a special heme protein—RHP (*Rhodospirillum* heme protein). This respiratory chain component has been intensively investigated and has now been obtained in pure form. Its magnetic and optical properties have been investigated (Ehrenberg and Kamen, 1965). It is present also in other photoheterotrophs—species of the genus Rhodopseudomonas (Bartsch and Kamen, 1958; Kamen and Vernon, 1955). In the structure of the prosthetic group, this heme protein (cytochromoid *C*) resembles cytochrome *c,* but it contains two heme groups per protein molecule. It reacts with carbon monoxide and, in the reduced form, with atmospheric oxygen. It is insensitive to cyanide.

Chance and Smith believe that RHP acts as the terminal oxidase in purple bacteria which lack type *A* cytochromes (Smith, 1959; Chance and Smith, 1955). Yet it has recently been shown that type *A* cytochrome can be found in a preparation of *R. spheroides* particles from cells grown aerobically in darkness. This oxidase can effect the aerobic oxidation of heart-muscle and yeast cytochrome *c* and bacterial cytochrome c_2. Cyanide inhibits it, while carbon monoxide in-

hibits it very weakly. A similar oxidase is found in *R. rubrum* cells
(Kikuchi *et al.,* 1965). These authors regard this enzyme as one of the
terminal oxidases of the dark respiration of nonsulfur purple bacteria. They suggest the following scheme of alternative pathways of
dark respiration for particulate preparations of the nonsulfur purple
bacterium *R. spheroides* (according to Kikuchi *et al.,* 1965)

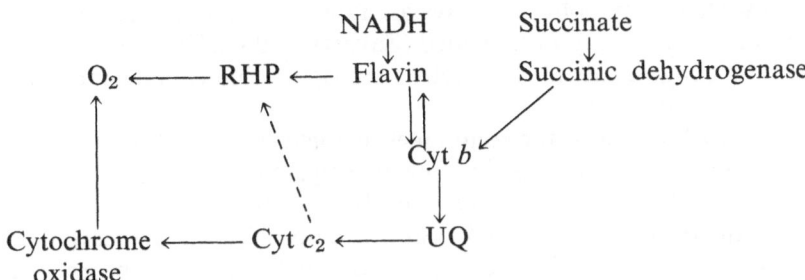

According to Taniguchi and Kamen (1965), cytochrome *o* and
the autoxidizable cytochrome *b* act as terminal oxidases in *R. rubrum*
cells grown aerobically in darkness. The authors suggest the following scheme of electron transport for *R. rubrum* in the exponential
phase of growth:

$$\text{Succinate} \longrightarrow \text{FP}_S \qquad \text{Exogenous } c_2$$
$$\searrow \qquad \nearrow$$
$$\text{UQ} - \text{Cyt } b \rightarrow \text{Cyt } o \rightarrow \text{O}_2$$
$$\nearrow \qquad \searrow$$
$$\text{NADH} \longrightarrow \text{FP}_N \qquad \qquad \text{Endogenous } c_2$$

FP$_S$ = succinic dehydrogenase
FP$_N$ = NADH dehydrogenase

Horio and Taylor (1965) suggest that RHP can be extracted from
cells in the form of cytochrome *o*. However, kinetic investigations of
the oxidase systems and the isolated protein of RHP from *R. spheroides* and *R. rubrum* raise doubt as to their identity (Chance *et al.,*
1966).

Chemoautotrophic bacteria (lithotrophs) obtain energy by oxidizing inorganic sulfur, iron, and nitrogen compounds, and molecu-

lar hydrogen. This group contains obligate aerobes (nitrifiers), as well as facultative (hydrogen bacteria, thiobacilli), and obligate anaerobes (sulfate-reducers). Comprehensive reviews of the metabolism, physiology, and energy metabolism of lithotrophic bacteria have been published (Zavarzin, 1964; Doman and Tikhonova, 1965; Ruban, 1961; Sokolova and Karavaiko, 1964; Nason, 1962; Peck, 1962; Lees, 1960; Schlegel, 1960; Postgate, 1959). Chemoautotrophic organisms with various metabolic pathways are now known, but only ten such organisms have been investigated (Zavarzin, 1964). The role of cytochrome components in the electron-transport chains of chemoautotrophs is of interest.

Cytochrome C is the commonest cytochrome among lithotrophs (see Table 11). It is found in all the investigated species. Cytochrome B, on the other hand, is often absent. In some lithotrophs (*Thiobacillus* X) the respiratory chain consists of a series of type C cytochromes. They differ in oxidation–reduction potential, but their spectra are identical (Trudinger, 1958, 1961).

According to Packer, the respiratory chain of hydrogen bacteria (*Hydrogenomonas*) contains cytochromes of types B and C and cytochrome oxidase. The terminal electron acceptor can be nitrate, as well as oxygen (Packer, 1958; Lees, 1958). The electron donor for the cytochromes can be either hydrogen or lactate, i.e., these bacteria can be either lithotrophs or organotrophs (Packer and Vishniac, 1955).

Lyalikova (1958) found cytochromes in the cells of the iron bacterium *T. ferrooxidans*. Cell-free preparations of *T. ferrooxidans* contain cytochromes of types C and A with absorption maxima for C at 552, 523, and 417 mμ, and for A at 590 and 439 mμ. Cytochrome c is present in very large amounts—1% of the dry weight of the bacteria (Vernon *et al.*, 1960). The cytochrome c was partially purified. Since it had been shown that the oxidation of Fe^{++} by the bacterial preparation is accompanied by reduction of the cytochromes, the authors proposed the following scheme for the respiratory chain (Vernon *et al.*, 1960).

$$Fe^{++} \rightarrow Cyt\ c \rightarrow Cyt\ a_1 \rightarrow O_2$$

It has been suggested that the initial enzyme of the electron-transport chain in this species of lithotroph is a specific iron oxidase localized on the outside of the cell wall. This enzyme leads to liberation of an electron from the bivalent iron ion. The electron is transferred to sulfate and then through a typical respiratory chain consisting of cytochromes c and, perhaps, UQ_6. The terminal electron acceptor is oxygen (Dugan and Lundgren, 1965).

Among the nitrifying bacteria, *Nitrosomonas* and *Nitrobacter* have been investigated. *Nitrosomonas* cells oxidize ammonia and hydroxylamine to nitrite in a respiratory chain which includes, in addition to flavin, cytochromes b, c, a_1, and a (Aleem *et al.*, 1963; Aleem and Lees, 1963; Falcone *et al.*, 1963; Anderson, 1964). *N. europea* contains two terminal oxidases: a cytochrome similar to the pigment P_{450} previously found in the microsomal fraction, and a soluble oxidase identified as cytochrome o (Rees and Nason, 1965). An enzyme complex (hydroxylamine:cytochrome c reductase) containing flavin, cytochromes b and c, and possibly iron has been extracted from *N. europea* and *N. oceanus* and purified (Hooper and Nason, 1965). Cytochrome c from *Nitrosomonas* has been isolated by Lozinov and Ermachenko (1962). Cytochromes are implicated in the oxidation of nitrite to nitrate in *Nitrobacter* (Lees and Simpson, 1957; Zavarzin, 1958). The cytochromes of this organism have been extracted and partially characterized (Butt and Lees, 1958). In Aleem and Nason's opinion (1960), the respiratory chain of *Nitrobacter* includes cytochromes c and a_1. Low-temperature spectrophotometry has shown the presence of cytochromes a, a_1, and c and the absence of cytochrome b in particles of *Nitrobacter winogradskii* (Van Gool and Laudelout, 1966). An enzyme complex containing cytochromes c, b, a, and a_1 has been isolated from *N. agilis*. It is inactivated by 5×10^{-5} M potassium cyanide and 2×10^{-3} M antimycin A (Straat and Nason, 1965).

The exact mechanism of oxidation of sulfur compounds by thiobacilli is unknown (Trudinger, 1964). There may be several pathways of conversion of sulfur (Vishniac and Trudinger, 1962; Suzuki, 1965). Three soluble type C cytochromes with α-bands at 550, 553, and 557 mμ have been isolated. A particulate preparation of this species was found to contain a cyanide-sensitive cytochrome oxidase, which could oxidize thiosulfate to tetrathionate with the aid of cytochrome c_{553} and the enzyme tetrathionase (Trudinger, 1958).

The presence of cytochromes in other thiobacilli (*T. denitrificans, Thiobacillus thioparus*, and *T. thiooxidans*) has been reported in several papers (Aubert *et al.*, 1959; Cook, 1964; London, 1963). The cytochrome c content of *T. denitrificans* reaches 4%. This cytochrome effects electron transfer between thiosulfate and nitrate in *T. deni-*

trificans (Aubert *et al.,* 1959). It has been shown for *T. thiooxidans* that cytochrome *c* takes part in the oxidation of sulfur compounds (London, 1963).

Obligate, anaerobic, sulfate-reducing bacteria (*Desulfovibrio*) have a very flexible metabolism. They can be strict lithotrophs, using hydrogen as an electron donor, and in this case assimilate CO_2. However, they can easily change over to organotrophy. In the absence of molecular hydrogen, they obtain energy by oxidizing lactate with sulfate. Cytochrome c_3 is involved in all these pathways (Postgate, 1956; Peck, 1962; Senez, 1962; Ishimoto and Yagi, 1961). This cytochrome has been isolated and is described above (Postgate, 1961). Cells and extracts of *D. desulfuricans* also contain a sulfate reductase which acts as the terminal oxidase of cytochrome c_3 (Egami *et al.,* 1961). Postgate notes that, despite its anaerobic nature, the metabolism of these bacteria is more like that of aerobes than that of fermentation anaerobes; it is more oxidative than fermentative (Postgate, 1961, 1965).

Some authors attribute the slow growth of lithotrophic organisms to the very small difference in the potentials of the initial and final components of the "shortened" chain, where, on the basis of thermodynamic considerations, only one pyrophosphate bond is formed (Newton and Kamen, 1963). This in turn leads to a greater synthesis of cytochromes than is the case in organotrophs (Vernon *et al.,* 1960; Aubert *et al.,* 1959; Ermachenko *et al.,* 1966; Van Gool and Laudelout, 1966).

ADDITIONAL COMPONENTS OF RESPIRATORY CHAIN

The mitochondrial respiratory chain includes nonheme iron, some of which is contained in the prosthetic groups of NADH and succinic dehydrogenases, and some of which is apparently in free form. The total nonheme iron content of mitochondria is 3.3 μmole per gram (Green and Fleischer, 1964).

It has been suggested that iron can undergo reversible reduction and oxidation (Green, 1962*a*, 1964*b*; Beinert and Lee, 1961; Doeg, 1961; Hatefi *et al.,* 1962).

Bacterial electron-transport particles also contain nonheme iron. However, the role of nonheme iron in bacterial electron transport has not been investigated. Hence, the hypotheses on this subject are based on an analogy with the mitochondrial process.

Succinate:cytochrome c reductase from *A. vinelandii* particles contains large amounts of nonheme iron in addition to other components, such as flavin, heme, and ubiquinone. The flavin:heme: nonheme iron ratio is 1:6:10 (Temperli and Wilson, 1962). A protein containing nonheme iron has been extracted from *A. vinelandii*. The authors regard it as a fragment of NADH dehydrogenase or succinic dehydrogenase, which is detached from these enzymes during extraction with butanol (Shethna *et al.*, 1966). Nonheme iron in amounts of 5–25 μmole per gram of dry weight has been found in various preparations containing the electron transport enzymes of *B. subtilis* (Downey, 1964).

A fraction of NADH oxidase particles from *M. tuberculosis* (strain H_{37}) contained 3.3–5.2 μmole per gram of nonheme iron protein. Its function was not investigated (Goldman *et al.*, 1963).

Nitrate reductase from *E. coli* contains one atom of molybdenum and about 40 atoms of nonheme iron per enzyme molecule. The best electron donor is cytochrome b_1 (Taniguchi and Itagaki, 1960). A similar enzyme has been obtained from *A. tumefaciens* (Kurup and Vaidyanathan, 1964).

It has been shown in two cases that nonheme iron is contained as the prosthetic group in purified flavoprotein dehydrogenases. The succinic dehydrogenase of *M. lactilyticus* contains 40 g atoms of nonheme iron per mole of FAD (Warringa *et al.*, 1958; Warringa and Giuditta, 1958). A purified preparation of succinic dehydrogenase from *pentosaceum* also contained large amounts of nonheme iron (Lara, 1959). It has been suggested that nonheme iron is connected with the enzyme activity of malate dehydrogenase from *Acetobacter xylinum* (Benziman and Galanter, 1964).

The copper content of mitochondria is 1.3 μmole per gram of protein. When electron-transport particles are obtained, the copper content is increased to 3.8 μmole per gram of protein (Green and Fleischer, 1964). It is believed that all the copper is contained in the cytochrome oxidase of heart-muscle mitochondria, in the amount of one or two molecules per enzyme molecule (Wainio, 1961).

Like nonheme iron, copper gives an electron-paramagnetic-resonance signal; it has been suggested that copper takes part in the oxidation–reduction reactions of the respiratory chain (Beinert *et al.*,

1962; Griffiths and Beinert, 1961; Griffiths and Wharton, 1961). It has been suggested that when cyanide acts on cytochrome oxidase, the cyanide forms complexes with copper as well as with heme (Frieden, 1964).

Some electron-transferring preparations of bacteria also contain copper. About 1 μmole of copper per gram of protein was found in a particulate preparation of NADH oxidase from *M. tuberculosis* (Goldman *et al.,* 1963). However, no copper was found in a purified preparation of bacterial cytochrome oxidase from *P. aeruginosa* (Yamanaka and Okunuki, 1963*b*). There is indirect evidence that copper is implicated in the cytochrome oxidase of the nitrifying bacteria *Nitrosomonas* and *Nitrobacter*.

Noncytochrome copper-containing proteins have been found in two species of *Pseudomonas* and are components of the respiratory chain (Horio *et al.,* 1961*a*; Coval *et al.,* 1961; Horio, 1958*b*; Ambler, 1963). This is the so-called blue protein, which contains one atom of copper per molecule. It has been purified and crystallized. Its molecular weight is 16,000–17,000, and the sequence of amino acids in its protein part has been determined. The protein contains 130 amino acids. In the oxidized form, blue protein has a characteristic maximum at 630 mμ. In the reduced form the maximum is absent. The bond between the copper and the protein can be broken by dialysis of the blue protein against a cyanide solution. The oxidation–reduction potential is $E'_0 = + 0.328$ v, and the isoelectric point is at pH 5.4. Blue protein reduced by lactic dehydrogenase or cytochrome c_{551} is easily oxidized by cytochrome oxidase from *P. aeruginosa,* but not by mitochondrial cytochrome oxidase (Horio *et al.,* 1961*a, b*). A similar blue protein, called azurin, has been extracted from *Bordetella pertussis, A. faecalis* and *Achromobacter denitrificans.* Its molecular weight is 14,000, and it also contains one copper atom per molecule. The blue protein is bound with cytochrome *c* and is reversibly oxidized by oxygen (Sutherland and Wilkinson, 1962). The information on copper-containing proteins has been summed up in a review (Frieden *et al.,* 1965).

LOCALIZATION OF RESPIRATORY-CHAIN ENZYMES

Having discussed the enzyme composition of the respiratory chain and its special features in various groups of bacteria, we now turn to the localization of respiratory-chain enzymes in the cell. The available information in this respect has been obtained by cytochemists and biochemists.

Cytochemistry of Respiratory-Chain Enzymes

As we showed in the first chapter, the cytoplasmic membrane and the membranous structures connected with it apparently carry various enzyme systems, the most important of which are the enzymes of the respiratory chain and oxidative phosphorylation. We noted the difficulties in determining the loci of these enzymes not only over the extent of any region of the membrane, but even in such large structures as the cytoplasmic membrane and mesosomes. Cytochemical data do not provide a conclusive answer to these questions, although they give some information.

Numerous cytochemical observations indicate that oxidation–reduction indicators are localized at the sites of the membranous structures. Fluorescent dyes such as berberin sulfate and aurophosphin have revealed granular formations in bacteria of the colon group (Meisel' and Mirolyubova, 1959). The use of oxidation–reduction indicators and electron microscopy has shown that in *E. coli* and *Sarcina ventriculi* the indicator deposits correspond in location to the membranous structures (Niklowitz, 1958). Janus green and formazan are also deposited in regions occupied by membranous structures in *E. coli* and *Sorangium compositum* (Biryuzova, 1960; Kushnarev, 1962). Later, Nermut showed that reduced tellurite is located in the form of aggregates on the periphery of the *P. vulgaris* cell (Nermut, 1960, 1963; Nermut and Rýc, 1964). Vanderwinkel and Murray (1962), using tetrazolium salts, which react with the respiratory chain at the level of cytochrome oxidase, showed that the oxidation–reduction enzymes are localized in the mesosomes of the aerobes *B. subtilis* and *S. serpens.*

Van Iterson and Leene (1964*a*) also found that in *B. subtilis* cells tellurite is deposited in the mesosomes, which they regard as equivalent to mitochondria. Since potassium tellurite was not concentrated in the cytoplasmic membrane, the authors rejected the view that the cytoplasmic membrane also contains the respiratory-enzyme chain. Similarly, formazan is deposited in the mesosomes of *B. mycoides* (Malatyan and Biryuzova, 1965). In this case the structure of the mesosomes is revealed. They consist of a system of tubules forming convoluted clumps without a bounding membrane.

These observations indicate that respiratory chain enzymes are absent from the cytoplasmic membrane, but do not agree with the results of other investigations. Mudd *et al.* (1961), who investigated the location of formazans in cells, protoplasts, and isolated membranes of *B. megaterium,* came to the conclusion that the cytoplasmic membrane and the mesosomes are a single functional system. Malatyan (1962), working with *E. coli,* found that redox indicators (fluorochromes, Janus green, and tetracycline) were concentrated equally in the membranous structures and the cytoplasmic membrane. In a later investigation, however, in which bromide-3-(4,5-dimethylthiazolyl-2)-2,5-diphenyltetrazolium was used on *E. coli* cells, no deposit of reduced indicator could be detected in the cytoplasmic membrane, and the formazan brought out only the membranous structures (Malatyan and Biryuzova, 1965).

Convincing data locating the oxidation–reduction enzymes in the cytoplasmic membrane of *E. coli* and *B. subtilis* are given by Sedar and Burde (1965*a, b*). The deposition of reduced tellurite in the cytoplasmic membrane of *L. monocytogenes* was distinct, whereas this indicator was not detected in the internal membranous structures. Formazan, on the other hand, was deposited only in the mesosomes (Takagi, Kawata *et al.,* 1963). The authors attribute this effect to the different set of dehydrogenases of the cytoplasmic membrane and mesosomes, but this is not very convincing, since the reduction of indicators is not a strictly specific process. It is more likely a case of different solubility of tellurite and formazan in the cytoplasmic membrane and mesosomes.

Yakovlev and Levchenko (1964) observed the deposition of formazans in the cytoplasmic membrane and in the membranous structures connected with it in *A. vinelandii.* In *P. vulgaris* and *S. aureus,* conglomerate deposits of tellurite were found near the cytoplasmic membrane, but no tellurite was found in the membrane itself (Van Iterson, 1965; Leene and Van Iterson, 1965; Suganuma, 1965).

Although cytochemical investigations using oxidation–reduction indicators to localize the respiratory enzymes have played a valuable part in establishing the connection between respiratory enzymes and the membranous structures of bacteria, cytochemical methods are not sophisticated enough to differentiate the loci of the respiratory chain. Bacterial cytochemistry has given conflicting answers even to such an apparently simple question as whether the respiratory chain is localized in the cytoplasmic membrane or in the mesosomes. The contradictions are probably due to the lack of strictly standardized staining and development techniques. According to Weibull (1953a), the picture of formazan distribution in *B. megaterium* cells changes in 10 min from diffusely scattered granules to distinct clumps. Sedar and Burde (1965a) and Takagi, Ueyama and Ueda (1963) also point out the need for strict control of conditions of work with tetrazolium indicators. We note in passing that an accumulation of reduced tetrazole has been observed in the mesosomes and cytoplasmic membrane of the obligate anaerobe *F. polymorphum*. It is obvious that the interaction of a tetrazolium derivative with dehydrogenases leads to its reduction, and formazan is then deposited in the lipid-rich membranes (Takagi, Ueyama and Ueda, 1963). These results indicate that oxidation–reduction indicators are not specific for the respiratory chain, because in anaerobes the respiratory chain is either reduced or absent altogether. If the development of mesosomes is assumed to be correlated with the respiration rate, the reason for the poor development of mesosomes in aerobic gram-negative heterotrophs is not clear. It seemed at one time that mesosomes could be revealed in gram-negative bacteria by the use of redox indicators.

According to recent results, however, these structures appear in response to the accumulation of formazan or tellurite (Leene and Van Iterson, 1965; Kushnarev, 1966). The loci of the respiratory chain enzymes in *P. vulgaris* are probably not the mesosomes, but membrane elements close to the cytoplasmic membrane.

Respiratory-Chain Enzymes in Structural Elements

Biochemical investigation of the localization of the respiratory-chain enzymes in bacteria did not become possible until methods of disintegrating and fractionating bacterial cells were devised. Several excellent reviews are available in which methods of mechanical, ultrasonic, and enzymatic disruption of cells are described and compared (Edebo, 1961; Hughes, 1961, 1962; Hughes and Cunningham, 1963; Salton, 1964). We will confine ourselves to the general comment that when any kind of mechanical disruption is used, the investigator will be dealing with a mixture of cell fragments, usually called particles or granules, which contain wall fragments, ribosomes, and fragments of the lipoprotein membranes. This mixture of intact and partially destroyed cell structures is very difficult to fractionate, although attempts have been made to do so (De Ley, 1962, 1963; Cota-Robles *et al.*, 1958; Schachman *et al.*, 1952; Weibull, 1953*c*; Tissières, 1961). Enzymatic digestion of the cell wall and extraction of the membranous elements in relatively intact form is a much more promising method, for example, by lysis with lysozyme, which is most suitable for work with gram-positive bacteria, but has been successfully used for work with gram-negative bacteria.

Investigation of the localization of respiratory-chain enzymes in particles was begun earlier. Weibull's experiments led to the development of research on enzymes in isolated membranes. In accordance with De Ley's suggestion (1962), particles carrying the respiratory-chain enzymes, as distinct from ribosomes, are sometimes called oxidosomes.

We think it worth while to discuss first the investigations of particles and then those of membranes. The investigation of enzymes in particles (oxidosomes) has been very fruitful, since many enzymes in various bacteria have been investigated. Since several reviews contain sections devoted to the localization of enzymes of the bacterial respiratory chain in particles, usually of 100–200 Å in diameter (Alexander, 1956; Bradfield, 1956; Gel'man, 1959; Marr, 1960; Mitchell, 1959*a, b*; Lascelles, 1965; Gel'man and Lukoyanova, 1962), we present only a summary table (Table 14).

Early in the investigation of the structure of the bacterial cell and the distribution of enzymes on the structural elements, it became clear that particles 100–200 Å in diameter could not in fact be the carriers of a complete respiratory chain. Calculations showed that a particle of such size could contain up to 14 protein molecules with a mean molecular weight of 35,000, whereas the mitochondrion contains up to a million such molecules (Bradfield, 1956). This gave rise to the suggestion that the particles result from the destruction of some larger structure. Weibull (1953*b*, 1956) succeeded in obtaining membranous elements of *B. megaterium* in an osmotically unstabilized medium by lysis of the cells with lysozyme. After Weibull's work, membranes were obtained from several bacteria, and their enzymatic composition was investigated but, unfortunately, by electron microscope in only a few cases. Information on the localization of respiratory-chain enzymes in membranes obtained from bacteria is given in Table 15.

It has not yet been possible to select conditions in which extracted mesosomes retain the same form as they have in the intact

Table 14. Respiratory-Chain Enzymes in Particles from Bacteria

Separation factor (g)	Enzyme	Bacterium	Author
25,000	Succinic oxidase	*Azotobacter vinelandii*	Nossal *et al.*, 1956
23,000	Cytochromes a_2, b_1, c_4, c_5 ⎫	*Azotobacter vinelandii*	Tissières, 1956;
145,000	Cytochromes c_4 oxidase ⎬		Tissières, *et al.*, 1957
144,000	Cytochromes a_2, b_1, c_4, c_5	*Azotobacter vinelandii*	Bruemmer *et al.*, 1957
144,000	NADH oxidase	*Azotobacter vinelandii*	Repaske and Josten, 1958
100,000	Cytochromes c_2, b, NADH oxidase, succinic oxidase	*Azotobacter agilis*	Cota-Robles *et al.*, 1958
20,000	Succinic oxidase ⎫	*Azotobacter vinelandii*	Hovenkamp,
147,000	Cytochromes a, a_2, b, c_4, c_5 ⎬		1959b
15,000	Succinic dehydrogenase, cytochromes a_1, a_2, b, c_4, c_5	*Azotobacter vinelandii*	Taniguchi and Ohmachi, 1960
144,000	Succinate:cytochrome c reductase, UQ_8	*Azotobacter vinelandii*	Temperli and Wilson, 1962
35,000	Flavin, UQ ⎫	*Azotobacter vinelandii*	Jones and
144,000	Cytochromes c_4, c_5, b_1, a_1, a_2 ⎬		Redfearn, 1966
40,000	Succinic oxidase, lactic oxidase, α-glycerophosphate oxidase	*Escherichia coli*	Asnis *et al.*, 1956
100,000	Cytochromes a_1, a_2, b_1	*Escherichia coli*	Tissières, 1961
15,000	Cytochrome b_1, Fe, Mo	*Escherichia coli*	Taniguchi and Itagaki, 1960
140,000	UQ_8 and $K_{2(45)}$, NADH oxidase	*Escherichia coli*	Kashket and Brodie, 1963a, b
78,000	Cytochromes a_1, a_2, b_1, UQ_8, K_2	*Escherichia coli*	Itagaki, 1964
100,000	Succinic oxidase	*Pseudomonas sp.*	Bentley and Schlechter, 1960
144,000	Flavin enzymes, cytochromes	*Pseudomonas sp.*	Stanier *et al.*, 1953
145,000	Cytochromes b, c_{553}	*Pseudomonas sp.*	Titani *et al.*, 1960
144,000	Malic oxidase	*Pseudomonas ovalis*	Kornberg and Phizack-erley, 1961

Table 14.   Continued

Separation factor (g)	Enzyme	Bacterium	Author
100,000	Lactic oxidase, cyto-chromes, ethanol oxi-dase	*Acetobacter peroxydans*	De Ley and Schell, 1959, 1962
105,000	Oxidases of primary, secondary, and poly alcohols	*Gluconobacter liquefaciens*	De Ley and Kersters, 1964; Kersters *et al.*, 1965
75,000	NADH oxidase, succinic dehydrogenase. Oxidation of glucose, gluconate, and 2, 5-diketo-gluconate	*Gluconobacter liquefaciens*	Stouthamer, 1961
110,000	Succinic oxidase	*Aerobacter aerogenes*	Tissières, 1954
38,000	NADH oxidase, succinic oxidase, malic oxidase, cytochromes of groups A, B, and C	*Mycobacterium avium*	Kusunose and Kusunose, 1959
140,000	NADH oxidase, malic oxidase, FAD, FMN, cytochromes a_3, a, b, c, c_1, vitamin $K_{2(45)}$	*Mycobacterium phlei*	Kashket and Brodie, 1960; Weber *et al.*, 1963; Brodie and Ballantine, 1960; Asano and Brodie, 1964, 1965a
105,000	NADH oxidase, cyto-chromes a_1, b, c_1	*Mycobacterium tuberculosis*	Bastarrachea and Goldman, 1961; Goldman *et al.*, 1963
100,000	NADH oxidase	*Mycobacterium avium*	Heinen *et al.*, 1964
25,000	Cytochrome b, malic oxidase	*Corynebacterium erythrogenes*	Tucker, 1960
105,000	Cytochromes a, a_3, b, c, succinic dehydrogenase	*Corynebacterium diphtheriae*	Pappenheimer *et al.*, 1962
105,000	Cytochrome $b_{(558)}$, succinic dehydrogenase	*Corynebacterium diphtheriae* PW 8 (Pd)	Pappenheimer *et al.*, 1962
144,000	Cytochromes, mena-quinones	*Corynebacterium diphtheriae*	Scholes and King, 1965b
100,000	Cytochromes a_1, a_2, b_1	*Xanthomonas phaseoli*	Hochster and Nozolillo, 1959

Table 14. Continued

Separation factor (g)	Enzyme	Bacterium	Author
100,000	Succinic oxidase, cytochromes a_1, a_2, b_1	*Xanthomonas phaseoli*	Madsen, 1960
130,000	NADH dehydrogenase	*Pasteurella tularensis*	Robinson and Mills, 1961
100,000	NADH dehydrogenase	*Proteus vulgaris*	Feldman and O'Kane, 1960
100,000	NADH oxidase, cytochromes	*Alcaligenes faecalis*	Pinchot, 1959
>20,000	Cytochromes a_{602}, b_{557}, b_{555}, succinic oxidase, lactic oxidase, α-glycerophosphate oxidase, NADH oxidase	*Staphylococcus aureus*	Taber and Morrison, 1964
144,000	NADH oxidase, cytochromes b_{560}, a_{610}	*Micrococcus lysodeikticus*	Ishikawa and Lehninger, 1962
20,000	NADH oxidase, cytochromes	*Micrococcus lysodeikticus*	Pardya and King, 1966
140,000	Cytochrome oxidase, NADH oxidase, cytochrome C	*Micrococcus denitrificans*	Vernon and White, 1957
70,000– 144,000	NADH:nitrate oxidoreductase; NADH:cytochrome c:O_2 oxidoreductase	*Micrococcus denitrificans, Pseudomonas denitrificans*	Nuik and Nicholas, 1966
144,000	NADH oxidase, cytochromes b, c, c_1, o, a_1, a_2, demethyl vitamin K_2	*Hemophilus parainfluenzae*	White and Smith, 1962; Smith and White, 1962; White 1962, 1963, 1965a, b
144,000	Nitrate reductase, cytochromes B, C	*Vibrio succinogenes*	Jacobs and Wolin, 1963
56,000	Cytochromes a_3, b, c	*Bacillus stearothermophylus*	Downey et al., 1962
100,000	NADH oxidase, cytochrome b_1	*Bacillus popilliae*	Pepper and Costilow, 1965
144,000	NADH oxidase, cytochromes a_1, b, c	*Bacillus cereus*	Doi and Halvorson, 1961
140,000	Succinic oxidase, NADH oxidase	*Bacillus anitratum*	Hauge, 1960
20,000	NADH oxidase, succinic oxidase, cytochromes	*Bacillus megaterium*	Pardya and King, 1966

Table 14. Continued

Separation factor (g)	Enzyme	Bacterium	Author
140,000	NADH oxidase, succinic oxidase, cytochromes a_1, a_2, b	*Achromobacter*	Mizushima and Arima, 1960
144,000	NADH oxidase, flavin, cytochrome c	*Hydrogenomonas euthropha*	Wittenberger and Repaske, 1961
145,000	Cytochromes $c_{547-549}$, $c_{553-555}$, cytochrome oxidase	*Thiobacillus X*	Trudinger, 1958, 1961
105,000	Fe^{++} oxidase, NADH oxidase, cytochrome c_{552}, $a_{588-590}$	*Thiobacillus ferrooxidans*	Vernon et al., 1960
45,000 144,000	Cytochromes a_{595}, b_{560}, c_{552}, Fe^{++} oxidase, cytochrome oxidase, cytochrome c reductase	*Ferrobacillus ferrooxidans*	Blaylock and Nason, 1963
144,000	NADH oxidase, NH_2OH: cytochrome c reductase, flavin, cytochromes b_{560}, c_{551}, a_{600}	*Nitrosomonas*	Falcone et al., 1962, 1963; Nicholas and Roa, 1964
100,000	NH_2OH oxidase	*Nitrosomonas*	Delvich et al., 1961
100,000	Cytochromes a, b, c	*Nitrosomonas europea, nitrosocystis oceanus*	Hooper and Nason, 1965
15,000	Cytochromes c_{550}, a_{598}	*Nitrobacter sp.*	Butt and Lees, 1958
27,000 58,000 95,000 144,000	Nitrite oxidase c_1, a_1, NADH oxidase	*Nitrobacter*	Aleem and Nason, 1959, 1960
40,000 120,000	NADH oxidase, nitrite oxidase, flavin, cytochromes c_{550}, a_{587}	*Nitrobacter*	Kiesow, 1964
150,000	Nitrite oxidase, cytochromes a, a_1, c	*Nitrobacter winogradskii*	Van Gool and Laudelout, 1966
15,000	Adenosinephosphosulfate reductase	*Desulfovibrio*	Ishimoto and Fujimoto, 1959
Particles obtained in sucrose gradient	NADH oxidase, succinic oxidase, cytochromes b, o_{564}, ubiquinone	*Rhodospirillum rubrum* (dark aerobic culture)	Taniguchi and Kamen, 1965
22,000	Succinic oxidase	*Rhodospirillum rubrum*	Smith, L., 1959
144,000	NADH oxidase, succinic oxidase	*Rhodopseudomonas spheroides* (dark aerobic culture)	Niederpruem and Doudoroff, 1965

Table 15.   Respiratory-Chain Enzymes in Bacterial Membranes

Enzyme	Bacterium	Author
Succinic dehydrogenase	*Bacillus megaterium*	Storck and Wachsman, 1957
Succinic oxidase	*Bacillus megaterium*	Nakada *et al.,* 1958
Succinic dehydrogenase, NADH oxidase	*Bacillus megaterium*	Weibull *et al.,* 1959
Cytochromes b, c	*Bacillus megaterium,* *Bacillus subtilis*	Vernon and Mangum, 1960
Cytochromes b_{560}, c	*Micrococcus lysodeikticus*	Jackson and Lawton, 1958, 1959
Cytochromes a_{601}, b_{560}, c_{550}, FAD, succinic dehydrogenase, NADH oxidase	*Micrococcus lysodeikticus*	Gel'man *et al.,* 1959, 1960a, b; Lukoyanova *et al.,* 1961; Gel'man, Zhukova, and Zaitseva, 1962
NADH oxidase, vitamin $K_{2(45)}$	*Micrococcus lysodeikticus*	Fujita *et al.,* 1966
Cytochromes a, b, c, succinic oxidase, NADH: cytochrome c reductase	*Vibrio cholerae* *Escherichia coli*	Murti, 1960
Cytochromes a, b, c	*Sarcina lutea*	Brown, J., 1961
Cytochromes, NADH oxidase	*Azotobacter agilis*	Robrisch and Marr, 1962
NADH oxidase, succinic oxidase, cytochromes c_{553}, b_{561}, malic oxidase	*Pseudomonas fluorescens*	Hughes *et al.,* 1959; Francis *et al.,* 1963
NADH oxidase	*Pseudomonas fluorescens*	Burrous and Wood, 1962
Glucose oxidase	*Pseudomonas aeruginosa*	Campbell *et al.,* 1962
D-alanine oxidase, cytochromes b, c	*Pseudomonas aeruginosa*	Norton *et al.,* 1963
Cytochromes, oxidation of glucose, D-lactate, gluconate, 2-ketogluconate	*Acetobacter* sp.	De Ley and Dochy, 1960a, b; De Ley and Kersters, 1964
NADH oxidase, succinic oxidase, flavoprotein, UQ_{10}, cytochrome, Fe_{NH}	*Agrobacterium tumefaciens*	Kurup *et al.,* 1966
Succinic dehydrogenase, lactic dehydrogenase, cytochrome b_5 reductase	*Escherichia coli*	Kidwai *et al.,* 1965
NADH oxidase, cytochromes b_1, a_1, a_2, o	*Escherichia coli*	Gray *et al.,* 1966

cell; the cytoplasmic membrane is better preserved (Beer, 1960; Lukoyanova *et al.,* 1961; Pangborn *et al.,* 1962). Since it has not yet been possible to separate the cytoplasmic membrane and mesosomes, in the discussion of the localization of the bacterial respiratory chain, we will speak of the membrane system. By this we mean the cytoplasmic membrane and the internal membranes.

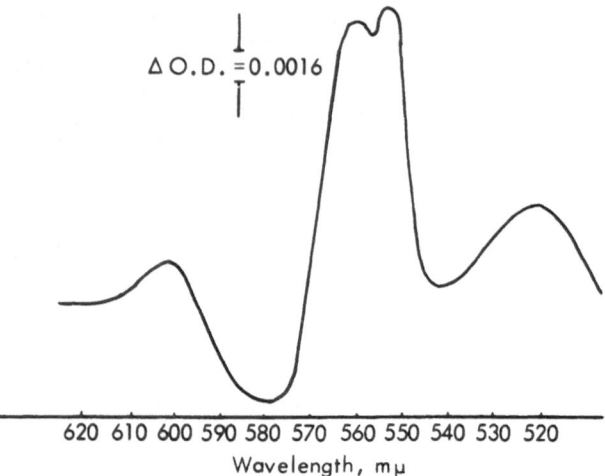

Δ O.D. = 0.0016

620 610 600 590 580 570 560 550 540 530 520
Wavelength, mμ

Fig. 16. Cytochrome membrane of *Micrococcus lysodeikticus*.

The main components of the chain, the cytochromes, are the most firmly bound and not generally found in the soluble part of the cell obtained after separation of the membranes (Fig. 16). The cytochromes of *P. aeruginosa, Thiobacillus X, N. winogradskii,* and *N. europea* are an exception: some are found in the soluble part (Eagon and Williams, 1959; Trudinger, 1961; Van Gool and Laudelout, 1966; Rees and Nason, 1965). The identity of the enzymatic activities of particles and membrane systems has been established in many investigations (Weibull, 1953*b, c*; Nakada *et al.,* 1958; Mitchell and Moyle, 1956; Mathews and Sistrom, 1959; Storck and Wachsman, 1957; Murti, 1960; De Ley and Dochy, 1960*a, b*; Stouthamer, 1961).

It is obvious that the membranes are the true carrier of the respiratory chain in bacterial cells and that the particles are obtained as a result of their destruction. This applies equally to gram-positive bacteria, where the membranous complex can be obtained in pure form, and to gram-negative bacteria, where it cannot always be separated from the cell-wall lipoprotein membrane, which, according to the data of several authors, is not implicated in cell respiration. It is of interest that in facultative anaerobes such as *E. coli* the respiratory chain enzymes remain firmly bound with the cell membrane even in anaerobic cultivation conditions (Gray *et al.,* 1966).

Attempts have been made to separate the cytoplasmic membrane from the mesosomal elements, but no clear results have yet been obtained. Pangborn *et al.* (1962) obtained *A. agilis* cell walls which contained cytoplasmic membranes and the network of internal membranes. Treatment in a disintegrator led to the release of the internal membranes and the simultaneous release of NADH oxidase. From these experiments the authors concluded that the respiratory chain is localized only in the internal membranes. In this connection we attempted to determine the localization of the respiratory-chain enzymes by lipase treatment of the membrane of *M. lysodeikticus* protoplasts in 1 *M* sucrose (Ostrovskii *et al.*, 1964).

The state of the protoplasts was checked by electron microscopy, from the turbidity and electrical conductivity of the suspension, and also from the protein and P_{inorg} content of the supernatant liquid after separation of the protoplasts. The functioning of the respiratory chain was determined by polarography (Ostrovskii and Gel'man, 1962; Ostrovskii, 1964a; Shol'ts and Ostrovskii, 1965).

Detectable lysis of the protoplasts (release of P_{inorg} and protein, destruction visible in electron microscope) did not occur until the respiration rate had fallen by 50–70% (Fig. 17). These data indicate that a considerable proportion of the respiratory-chain enzymes is concentrated in the cytoplasmic membrane of the cell. This is also indicated by simple calculations. According to Lehninger

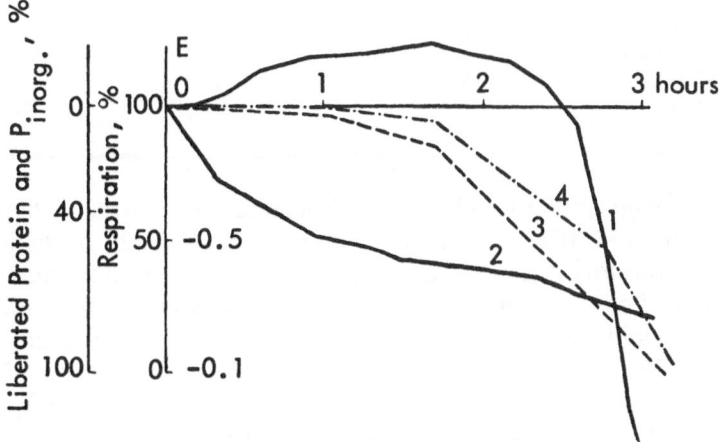

Fig. 17. Effect of lipase on protoplasts of *Micrococcus lysodeikticus*. (1) Optical density of lipase-treated suspension in comparison with control suspension; (2) respiration rate of lipase-treated protoplasts, percent of respiration rate of control protoplasts; (3) amount of dissolved protein, percent of amount in total lysis; (4) amount of P_{inorg}, percent of amount in total lysis.

Table 16. Amounts of Respiratory-Chain Components in Membranes or Fragments (in μmole per g of protein)

Bacterium	FAD+ FMN	Naphtho- quinones	Ubiqui- nones	Cytochromes			Fe^*_{NH}	Author
				A	B	C		
Bacillus cereus	0.40	—	—	0.19	0.39	0.18	—	Doi and Halvorson, 1961
Micrococcus lysodeikticus	—	2.5	—	0.17	0.32	0.42	—	Lukoyanova, 1964; Fujita et al., 1966
Mycobacterium phlei	0.68	12.0	—	0.27	0.18	0.62	—	Asano and Brodie, 1964; Kashket and Brodie, 1960
Rhodospirillum rubrum	0.60	—	0.30	—	0.24 (b_{564}) 0.30– 0.46 (b)	0.1– 0.15	0.80	Taniguchi and Kamen, 1965
Escherichia coli	0.12	0.86	1.26	0.1	0.62	—	—	Itagaki, 1964
Nitrosomonas	0.30	—	—	0.50	0.34	3.82	—	Falcone et al., 1962, 1963
Agrobacterium tumefaciens	0.31	—	2.4	—	—	0.3	7.14	Kurup et al., 1966
Azotobacter vinelandii (small particles)	2.0	None	8.9	0.67	1.59	1.62	—	Jones and Redfearn, 1966
Hemophilus parainfluenzae	—	0.61	—	—	—	—	—	White, 1966

(1964), the molecular weight of the mitochondrial respiratory chain is 1.3–1.8 million. If we assume that the molecular weight of the bacterial respiratory chain is also of the order of 1.3 million, and the quantity of chain components is about 0.2 μmole per gram of membrane protein (Table 16), then the weight of respiratory chain enzymes is 26% of the weight of the protein of all the membranes of *M. lysodeikticus*. We assume here that the relative amounts of respiratory-chain enzymes and other proteins in bacterial membranes are the same as in mitochondria. Yet the surface of the internal membranes is only 30% of the surface of all the cell membranes (cytoplasmic and internal membranes), based on estimates from electron micrographs of bacterial sections (Salton and Chapman, 1962). Thus, the assumption that the respiratory chain is localized only in the internal membranes (mesosomes) leads to the assumption that almost 80% of the mesosome proteins consists of enzymes, which is quite improbable, since even for proteins of mitochondrial membranes the enzyme content of the respiratory chain does not exceed 25% (Lehninger, 1964). This contradiction is removed if it is assumed that about 60% of the respiratory chain enzymes is localized in the cytoplasmic membrane of *M. lysodeikticus.*

Shah and King (1965) give very interesting data on the localization of the respiratory chain in *M. lysodeikticus.* Succinate oxidation in the succinic-oxidase system is inhibited when the cell is lysed with lysozyme, but is hardly inhibited at all if ultrasound is used to disrupt the cell. If the ultrasonic fragments are then treated with lysozyme, succinic oxidase is inhibited. On the other hand, the activity of NADH oxidase can even be increased by the treatment of ultrasonic fragments with lysozyme. Assuming that mucopeptide, the substrate on which lysozyme acts, is found only in the cell wall, these experiments can be interpreted in only one way, *viz.,* that the respiratory chain and particularly succinic oxidase are connected in some way with the cell wall in view of their localization in the cytoplasmic membrane.

As Tables 14 and 15 show, the main components of the respiratory chain—succinic dehydrogenase, NADH dehydrogenase, cytochromes of groups *A, B,* and *C,* ubiquinones, and naphthoquinones —are found in membranes or in particles obtained by homogenization and precipitation in a fairly wide range of accelerations. Thus, in composition of respiratory-chain components, bacterial membranes are very similar to mitochóndria.

The amounts of cytochromes and other respiratory-chain components in bacterial membranes or particles have been determined in only a few investigations (Table 16). Unfortunately, it is difficult to estimate the range of variation of the amounts of respiratory-chain components in bacterial membranes, since only a few species have been investigated, and the choice of organisms has been arbitrary. The flavin content is comparable with its content in mammalian mitochondria and electron-transport particles—0.36–0.60 μmole per gram of protein (Green and Wharton, 1963; King *et al.,* 1964). Again, as in mammalian mitochondria, ubiquinones and menaquinones are in great excess in comparison with the other respiratory-chain components.

Table 17.   Firmly Bound Dehydrogenases in Bacterial Particles and Membranes

	Enzyme	Prosthetic group	Bacterium	Author
	Particles			
29S and 40S	Formic dehydrogenase	?	*Escherichia coli, Pseudomonas fluorescens*	Schachman *et al.*, 1952
38,000g	Malic dehydrogenase	?	*Mycobacterium avium, Mycobacterium stegmatis, Mycobacterium butiricum*	Kusunose and Kusunose, 1959
78,000g	Malic dehydrogenase		*Micrococcus lysodeikticus*	Cohn, 1956, 1958
105,000g	Dehydrogenases of primary, secondary and poly alcohols, pentoses, aldoses	Cytochrome c_{553}?	*Gluconobacter suboxydans*	De Ley and Kersters, 1964; Kersters *et al.*, 1965
100,000g	Lactic dehydrogenases	?	*Acetobacter peroxydans*	De Ley and Schell, 1959; Iwasaki, 1960
105,000g	Glucose dehydrogenase, ethanol dehydrogenase	?	*Acetobacter peroxydans*	King and Cheldelin, 1957, 1958
90,000g	Glucose-6-phosphate dehydrogenase, 6-phosphogluconate dehydrogenase	NAD	*Erwinia amylovorum*	Sutton and Starr, 1960
80,000g	Malic dehydrogenase	Lipid, FAD	*Mycobacterium avium*	Kimura and Tobari, 1963; Tobari, 1964
30,000g	Alcohol:cytochrome c_{553} reductase, aldehyde dehydrogenase	Cytochrome c_{553}	*Acetobacter sp.*	Nakayama, 1961; Nakayama and De Ley, 1965
140,000g	Aldose dehydrogenase	?	*Pseudomonas fragi*	Weimberg, 1963
100,000g	D-allohydroxyproline dehydrogenase	?	*Pseudomonas striata*	Adams and Newberry, 1961
140,000g	Aldehyde dehydrogenase	?	*Pseudomonas aeruginosa*	Heydeman and Azoulay, 1963
140,000g	Malic dehydrogenase	FAD?	*Pseudomonas sp.*	Kornberg and Phizackerley, 1961; Francis *et al.*, 1963

Table 17. Continued

	Enzyme	Prosthetic group	Bacterium	Author
		Particles		
140,000g	Glucose dehydrogenase	X(maximum 337 mμ)	*Acetobacter suboxydans, Pseudomonas fluorescens*	Hauge, 1961a, b, 1964; Hauge and Hallberg, 1964; Hauge and Mürer, 1964
144,000g	Steroid-1-dehydrogenase	FAD?	*Nocardia*	Sih and Bennet, 1962
105,000g	D-Fructose dehydrogenase	?	*Gluconobacter acinus*	Yamada et al., 1966
144,000g	D- and L-lactic dehydrogenases, L-malic dehydrogenase	?	*Azotobacter vinelandii*	Jones and Redfearn, 1966
140,000g	Lactose dehydrogenase	FAD	*Pseudomonas graveolens*	Nishizuka and Hayashi, 1962
144,000g	Aldose dehydrogenase	?	*Rhodopseudomonas spheroides*	Niederpruem and Doudoroff, 1965
123,000g	D- and L-lactic dehydrogenases	?	*Escherichia coli*	Kline and Mahler, 1965
		Membranes		
	Lactic dehydrogenases	?	*Pseudomonas natriegenes*	Walter and Eagon, 1964
	Lactic dehydrogenase, malic dehydrogenase	?	*Bacillus megaterium*	Weibull et al., 1959
	D-alanine dehydrogenase	?	*Pseudomonas aeruginosa*	Norton et al., 1963
	D-lactic dehydrogenase, L-lactic dehydrogenase, formic dehydrogenase	?	*Hemophilus parainfluenzae*	White, 1964
	Malic dehydrogenase	? / FAD	*Micrococcus lysodeikticus*	Gel'man et al., 1960a, b, 1963
	Trimethylamine-N-oxidoreductase	?	*Vibrio parahaemolyticus*	Unemoto et al., 1965

In almost all the investigated bacteria except *Azotobacter,* the membranes are poorer in cytochromes than heart mitochondria, but are comparable with the mitochondria of other organs. For instance, rat-liver mitochondria contain 0.84 μmole of cytochromes per gram of protein (Estabrook and Holowinsky, 1961). Chemoautotrophs are exceptionally rich in cytochromes, a feature associated with the special nature of their metabolism. Several papers give comparative data on the cytochrome content of particulate preparations in arbitrary units. Such data have been calculated for *H. parainfluenzae, B. megaterium, M. lysodeikticus,* and *E. coli* (White, 1966; Pandya and King, 1966; Gray *et al.,* 1966).

In addition to the respiratory-chain components listed, the membranes and their fragments contain firmly bound dehydrogenases, a list of which is given in Table 17. The strength of the bond of the dehydrogenases with the membranes varies. According to White (1965*b*), sonication of *H. parainfluenzae* cells leads to the solution of 29% of the succinic dehydrogenase, 36% of the NADH dehydrogenase, and 92% of the lactic dehydrogenase. In the "detached" state, the dehydrogenases are incapable of interacting with the respiratory chain.

ROLE OF LIPIDS IN ORGANIZATION OF RESPIRATORY CHAIN

It has been reported on many occasions that the membrane system of bacteria (cytoplasmic membrane and internal membranous structures) contains up to 30% lipids, mainly phospholipids. Bacterial membranes are similar to mitochondrial membranes in this respect. Intensive research on the role of lipids in the organization and action of respiratory chain enzymes in mitochondrial membranes has been conducted in the last ten years, but research on bacterial membranes has just begun. There are numerous observations which show that destruction of the lipids in mitochondrial membranes by bile salts, detergents, and lipolytic enzymes, or extraction of the lipids leads to inactivation of the respiratory chain. Inactivation due to extraction of the lipids with a solvent may be reversible in certain conditions and the activity restored by replacement of the lipid. There is no clear-cut theoretical interpretation of these experimental results, although the responsible role of lipids in electron transport in the mitochondrial membrane is beyond question. An obvious first hypothesis on the role of lipids in electron transport is that the lipids are implicated in structure formation—in the assembly of the membrane. Hence, their removal or a change in their composition and structure impairs the assembly of the respiratory-chain components and inhibits electron transport. We can give several examples which illustrate the inhibition of the mitochondrial respiratory chain due to alteration of the lipids. Inactivation of succinic oxidase in mitochondrial preparations by detergents has been observed by several authors (Hockenhull, 1948; Okui *et al.*, 1963; Wills, 1954). Slater (1959) described the inhibition of mitochondrial NADH oxidase by sodium desoxycholate. Destruction of the mitochondrial structure and the simultaneous inhibition of the respiratory chain due to phospholipase A or C have been observed (Edwards and Ball, 1954; Petruschka *et al.*, 1959).

It has been shown that the accomplishment of electron transport in mitochondria requires the presence of phospholipids (Fleischer *et al.*, 1952). This process is completely suppressed if the phospholipids are removed with aqueous acetone and is restored only by replacement of the phospholipids in solubilized micellar form (Green and Fleischer, 1963). The reversibility of this process suggested to Green and his colleagues that phospholipids, situated between the interacting links of the respiratory chain, form a nonaqueous medium in which oxidation–reduction reactions occur. This is particularly important for oxidative phosphorylation, since it is believed that the primary complexes associated with the formation of the high-energy bond require protection from the hydrolytic action of water. However, the hypothesis that a lipid medium is necessary for electron transport has been criticized. Some authors believe that the process is more likely to occur in an aqueous medium (Chance, 1964; Tyler and Estabrook, 1966).

The whole collection of information indicating inhibition of the respiratory chain due to damage to lipids of the mitochondrial membranes or their fragments does not imply that the lipids play a functional role, since in such experiments it is very difficult to separate the functional and structural role of lipids. Investigation of the role of lipids in the activity of individual respiratory enzymes are required to elucidate the role of lipids in the function of the respiratory chain. That lecithin plays a part in the activity of β-hydroxybutyric dehydrogenase has been established (Jurtshuk *et al.*, 1961, 1963). Preparations of purified mitochondrial cytochrome oxidase also regularly contain 10-20% phospholipids (Cohen and Wainio, 1963; Brierley and Merola, 1962). According to Lemberg *et al.*, 1962), their role is to form a bond between heme and protein. Other authors think that the phospholipids of cytochrome oxidase are responsible for electrostatic binding of this enzyme with cytochrome c (Matsubara *et al.*, 1965; Tzagoloff and MacLennan, 1965). Specific activation by phospholipids has been shown in the case of purified mitochondrial succinic dehydrogenase (Cerletti *et al.*, 1965).

It has been shown that NADH:ubiquinone reductase contains about 20% lipids. If they are destroyed by treatment with solvents or phospholipase A the enzymatic complex is inactivated (Redfearn and King, 1964; Fleischer, 1964; Minakami *et al.*, 1964). Isolated mitochondrial cytochromes contain lipids. For instance, cytochrome b contains 10% lipids. Although cytochrome c can be obtained in lipid-free form, it can easily form a stable, highly active complex with phosphatidylethanolamine (Das and Crane, 1964; Das *et al.*, 1964).

Pesch and Peterson (1965) have shown that the removal of lecithin from mitochondrial NADPH:NAD transhydrogenase alters the substrate specificity of the enzyme.

Experimental data on the role of lipids in the activity of individual enzymes obviously provide a better basis for assessment of the function of lipids in the electron transport chain than experiments with intact membranes. It is believed that complexing the enzyme protein with phospholipid ensures the conformation necessary for the enzyme activity.

Investigations of the role of lipids in the spatial organization of the respiratory chain in bacterial membranes give results which are basically similar to those obtained for mitochondria.

Destruction of mycobacterial membranes with detergents leads to inactivation of the respiratory chain (Goldman *et al.*, 1963). Iwasaki showed the inhibiting effect of Emasol and Tween 80 on the lactic oxidase system of *A. suboxydans* (Iwasaki, 1960).

In a concentration of 0.01% Triton X-100 caused 70–80% inactivation of the succinic oxidase of a cell-free preparation of *C. diphtheriae*. The reduction of cytochromes c and a was simultaneously inhibited. NADH oxidase, however, was not affected in this case (Scholes and King, 1965*a*).

Destruction of membranes of *M. lysodeikticus* with desoxycholate and Tween 80 was accompanied by the inactivation of succinic oxidase, malic oxidase, and NADH oxidase (Gel'man *et al.*, 1960*b*; Lukoyanova *et al.*, 1961).

It has been shown that the removal of lipids from various preparations of bacterial membranes containing the respiratory chain inhibits electron transport to oxygen. The lipids are extracted as in work with mitochondria—in the cold, for several minutes. To judge from the difficulty of extracting lipids with cold solvent, the bond of lipids with protein in bacteria is stronger than in mitochondria. This may be attributable to the different composition of the phospholipids. In the opinion of Strickland and Benson (1960), phosphatidylglycerols (which are so common in bacterial membranes) have the ability to form stable complexes with proteins both by ionic and hydrogen bonds and by Van der Waals forces. Unfortunately, the methods of extracting bacterial particles with solvents have no theoretical basis. The nature of the solvent and the completeness of extraction of the lipids are not always taken into consideration. This probably accounts for the ease of reconstructing the respiratory chain with ubiquinones or naphthoquinones alone, which are more easily extracted with solvents than lipids bound in lipoprotein complexes. Hence, the results obtained by means of this technique cannot be regarded as conclusive in relation to the role of lipids in bacterial electron transport. In several cases it is difficult to determine if there is a specific requirement for the lipid fraction or ubiquinones and naphthoquinones. Kashket and Brodie (1963a) showed that treatment of electron-transport particles of *E. coli* with isooctane suppressed the oxidation of succinate and malate. Reactivation was achieved by the addition of vitamin K_2 and the phospholipoprotein material which was removed during extraction. Similar results were obtained for *B. stearothermophilus* (Downey *et al.*, 1962). Extraction with organic solvents (ethanol-ether, acetone) destroyed 90% of the succinic oxidase activity of the particles. The initial activity was restored by the simultaneous addition of vitamin K_1 or UQ, but lipid extract was also required. Isobutanol treatment of *E. coli* particles carrying the respiratory chain inhibited electron transport from NADH to nitrate. Unfortunately, the nature of the active principle was not determined (Taniguchi *et al.*, 1958; Iida and Taniguchi, 1959). In the majority of investigations in which lipids have been extracted, the process could be reliably reconstructed only by quinones (Van Demark and Smith, 1964; Itagaki, 1964; Segel and Goldman, 1963).

More definite information on the role of lipids in the respiratory chain can apparently be obtained by the use of lipolytic enzymes, which cleave lipids without denaturing protein or attacking quinones. Treatment of *E. coli* particles with snake venom led to the extraction of a complex of formic dehydrogenase, cytochrome *b*, and nitrate reductase (Itagaki *et al.*, 1961). Nitrate reductase and cytochrome oxidase can be solubilized from particles of a halotolerant micrococcus by combined treatment with lipase, snake venom, and desoxycholate (Hori, 1963).

We also attempted to determine the role of lipids in electron transport. To investigate the function of lipids, we selected a specific treatment for membrane lipids of *M. lysodeikticus,* treatment with lipolytic enzymes: pancreatic lipase and phospholipase from cobra venom (Ostrovskii *et al.,* 1964; Ostrovskii, 1964*b*; Lukoyanova, 1964). Our purpose was to find how the integrity of the phospholipids and glycerides affected the activity of the whole respiratory chain. Treatment of the membranes with lipolytic enzymes led to 80–90% inhibition of the electron-transport chain (Table 18), which agrees with the data on the inhibiting effect of lipase on *M. lysodeikticus* protoplasts (Ostrovskii *et al.,* 1964).

It is reported in the literature on mitochondria that free fatty acids inhibit some oxidases (Dalgarno and Birt, 1963). Fatty acids in a concentration of 1 μmole of fatty acid (C_{10}–C_{14}) apparently have a solvent effect, like detergents, on mitochondrial membranes (Baker *et al.,* 1962). The addition of serum albumin to the medium usually helps to protect respiration from the harmful effect of fatty acids, and a stable complex with the fatty acid is formed (Borst *et al.,* 1962). We

Table 18. Inhibition of Oxidases by Lipolytic Enzymes (in μmole of oxidized substrate per mg of protein per hr)

Lipolytic enzyme	Malic oxidase		NADH oxidase	
	control	experiment	control	experiment
Lipase	12.0	2.4	9.6	0.96
Phospholipase A	9.75	1.16	—	—

The membranes were incubated with lipase for 60 min, and with phospholipase for 180 min, at 37°C and pH 7.4. The concentration of membrane protein was 7.5 mg/ml.

The malic oxidase activity was measured manometrically at 30°C. For the activity measurements 2 ml of 0.1 M Na-K phosphate buffer (pH 7.4) and 0.5 ml of membrane suspension (2–4 mg of protein) were put into the Warburg vessel; 25 μmole of potassium malate was put into the side vessel; 0.2 ml of 33% NaOH solution was put into the control vessel.

The NADH oxidase activity was determined spectrophotometrically by the reduction of optical density at 340 mμ during oxidation at 30°C. The experimental cuvette contained 3 ml of 0.01 M phosphate buffer (pH 7.4), membrane suspension in an amount of 100–200 μg of protein, and substrate in an amount of 0.5 μmole of NADH.

added 2 mg BSA per milligram of membrane protein to the sample at the same time as the lipase on the assumption that the fatty acids, as soon as they were liberated from the membrane (if this was the mechanism of inhibition), would be adsorbed by the albumin. However, no reduction was noted in the inhibition of the respiratory chain.

The data in Figs. 13 and 14 also indicate that treatment with lipolytic enzymes for 1–3 hr did not cause the membranous structures to break up into particles, although there was a change in the ultrastructure of the membranes—perforations, stratification, and disappearance of the subunits (Lukoyanova and Biryuzova, 1965).

It was of interest to determine the points of injury to the respiratory chain by lipolytic enzymes. We began our investigation of the mode of injury with dehydrogenases. The malic dehydrogenase of the membranes was inhibited 50% on the average by both lipolytic enzymes. We can infer that the first link in the chain of malate oxidation makes a contribution to the inhibition of malic oxidase due to lipase or phospholipase. The succinic dehydrogenase activity was inhibited almost completely by lipase, whereas phospholipase had no effect on this dehydrogenase. NADH dehydrogenase was not affected at all by lipase. The acceptors were: 2,6-dichlorophenolindophenol for malic dehydrogenase and NADH dehydrogenase, and phenazine methosulfate and 2,6-dichlorophenolindophenol, according to Ells' method (1959), for succinic dehydrogenase. We then investigated the relationship between malic dehydrogenase and NADH dehydrogenase in an enzyme preparation, the preparation and properties of which we described in the section on respiratory-chain dehydrogenases. The enzyme activity in the purified preparation still depended on the phospholipids and glycerides. Malic dehydrogenase was 80% inactivated by treatment of the preparation with lipase or phospholipase; NADH dehydrogenase was only 15–20% inactivated. In the case of malic dehydrogenase, phospholipids and glycerides apparently form a complex which ensures the conformation necessary for the activity, while for NADH dehydrogenase, complex formation is only a means of attaching the enzyme to the respiratory chain (Oparin et al., 1965; Deborin et al., 1966).

The dehydrogenases of M. lysodeikticus membranes are no exception as regards binding with lipids. The malic dehydrogenase of M. avium particles requires diphosphatidylglycerol for its action (Tobari, 1964; Kimura and Tobari, 1963). The soluble malate:vitamin K reductase of *M. phlei* is activated by small amounts of various phospholipids. The enzyme activity is suppressed by Tween 80, desoxycholate, and Triton X-100 (Asano *et al.,* 1965).

From these data we can postulate that the lipids in bacterial membranes determine the spatial disposition and activity of some dehydrogenases. There are similar data for mitochondrial enzymes (succinic dehydrogenase, β-hydroxybutyric dehydrogenase).

It seems quite likely that alteration of the native state of the lipid components of *M. lysodeikticus* membranes can partially solubilize the cytochromes, so that oxidation in the respiratory chain decreases. It is known that mitochondrial cytochrome *c* can be brought into solution from a complex with lipid by phospholipase A (Ambe and Crane, 1959). Yet we did not observe solubilization of cytochromes during the action of lipolytic enzymes in our experiments.

Interesting results in this connection are those of Hori, who found that the cytochrome oxidase of a halotolerant micrococcus could be solubilized only by the combined action of lipase, phospholipase A, and desoxycholate (Hori, 1963).

It could be postulated that suppression of electron transport in the membrane would be accompanied by appreciable destruction of the lipid components. In this connection we investigated the chemical aspect of the changes produced by lipolytic enzymes (Oparin, Lukoyanova *et al.,* 1965). Infrared spectroscopy revealed a reduction in the amount of ester groups in the lipid fractions of the membranes after treatment with phospholipase and lipase by 10 and 19% relative to a control. On conversion, the amount of liberated fatty acids was 0.02 μmole/50 ml of suspension containing 400 mg of membrane protein. It is evident that slight hydrolysis of the membrane lipids leads to reorganization of the lipoprotein complexes and inhibition of the respiratory chain. Thus, one of the functions of phospholipids and glycerides in bacterial membranes carrying the respiratory chain is to create a structure which determines the spatial disposition of the enzymes in the multicomponent system. One of the finer mechanisms may be that these lipid components form complexes with individual enzymes, thus determining their activity (malic dehydrogenase, succinic dehydrogenase) or their mode of connection with the respiratory chain (NADH dehydrogenase).

Inhibition of electron transport by lipolytic enzymes also occurs in mitochondria (Edwards and Ball, 1954; Minakami *et al.,* 1964;

Hatefi *et al.,* 1962; Petruschka *et al.,* 1959; Green and Fleischer, 1964). However, in bacterial membranes, as distinct from mitochondria, glycerides as well as phospholipids are implicated in the maintenance of the membrane structure and in the function of the respiratory chain. Downey *et al.,* (1962) reported that the size of electron-transport particles from *B. stearothermophilus* was reduced by the action of lipase.

FUNCTIONS OF SUBUNITS

The function of the subunits is a topic of lively discussion. Their discovery on the membranes, where electron transport is effected, suggests their connection with this process.

According to some authors (Green, 1964*a, b*; Blair *et al.,* 1963; Fernandez-Móran, 1963; Fernandez-Móran *et al.,* 1964), each mitochondrial subunit contains a complete respiratory chain. These authors cite evidence to show that the subunits are identical with the elementary particles previously obtained by fractionation of the mitochondria. This view, however, is not generally accepted. Racker, Chance, and Parsons found a large number of subunits in mitochondrial membranes capable of oxidative phosphorylation. Membranes lacking coupling factors contained few subunits, and the addition of factors which restored oxidative phosphorylation led to an increase in the number of subunits (Racker *et al.,* 1964; Kagawa and Racker, 1965). The subunits contained ATPase (Racker, 1965). Direct evidence of absence of respiratory-chain components in the subunits has been obtained by several authors, who showed cytochromes localized in membranes from which the subunits had been removed (Williams and Parsons, 1964; Chance, 1965).

Other authors are also doubtful that the subunits are loci of the respiratory chain (Stasny and Crane, 1964; Lusena and Dass, 1963; Greville *et al.,* 1964; Chance and Parsons, 1963).

Above it was shown that lipolysis causes the subunits to disappear from *M. lysodeikticus* membranes (see Figs. 13 and 14). It would appear that the disappearance of the subunits and the concomitant inhibition of the electron transport chain in these experiments could be interpreted as evidence that the respiratory chain is indeed localized in the subunits. According to our results, however, the disappearance of the subunits did not lead to the liberation of cytochromes. This suggests that the structure carrying the main components of the electron-transport chain is the membrane, not the subunits. The fate of the subunits could not be followed and, hence, it is quite possible that the disappearance of the subunits is merely a consequence of alteration of the membrane ultrastructure by the action of lipolytic enzymes. This may cause the subunits to fuse with the carrier membrane, and the enzymes localized in the subunits will remain bound with the membrane. The disappearance of the subunits as a result of lipolysis is consistent with the report of their destruction by treatment with desoxycholate for more than 8 min (Lusena and Dass, 1963). This suggests that lipids are implicated either in the construction of the subunits or their attachment to the membrane.

It is apparently still too soon to come to a definite conclusion regarding the function of subunits in bacterial and mitochondrial electron transport. The solution of this question is impeded by the lack of reliable techniques for separation of the subunits from the membranes.

Another point is important. Subunits similar to mitochondrial subunits are found on the internal and cytoplasmic membranes of nine investigated species of bacteria in addition to *M. lysodeikticus: Bacillus pumilis, B. licheniformis, Bacillus brevis, Bacillus circulans, E. coli* C, *P. vulgaris, Shigella dysenteriae* Y6R, and *B. stearothermophilus* (Abram, 1965). This indicates the functional similarity of bacterial and mitochondrial membranes, although it is still impossible to define precisely the role of the subunits. Abram's information makes it easier to understand the relative importance of the cytoplasmic membrane and bacterial membranes in the organization of the bacterial respiratory chain. Since subunits are a characteristic feature of membranes carrying the respiratory chain and are present on the cytoplasmic membrane, the latter also must carry the respiratory chain. This is perfectly consistent with experimental data indicating that 60–70% of the respiratory chain is localized in the cytoplasmic membrane (Ostrovskii *et al.*, 1964).

EFFECT OF DEFORMATION OF MEMBRANE ON OPERATION OF RESPIRATORY CHAIN

Since a change in the chemical composition of membrane lipids leads to a change in the ultrastructure of the membrane, it is not surprising that it also will severely damage the enzyme systems mounted on the membranous structures. Yet other kinds of treatment, which probably cause only deformation of the membrane, can also injure the respiratory chain. Such treatments include the rupture of the cell and the associated uptake of water and swelling of the membrane, and in some cases mechanical injury to the latter. The respiration rate of protoplasts obtained in osmotically stabilized solutions usually does not differ from that of intact cells. In protoplasts of *A. peroxydans* and *Gluconobacter liquefaciens,* numerous substrates are oxidized at almost the same rate as in intact cells (De Ley and Dochy, 1960*a, b*). Osmotic shock of protoplasts usually leads to some inhibition of oxidation. Jose and Wilson (1959) observed much slower succinate and malate oxidation by *A. vinelandii* membranes as compared with protoplasts. Nermut also observed inhibition of amino-acid oxidation, measured by the reduction of tellurite, when penicillin spheroplasts of *P. vulgaris* were lysed. At the same time, oxidation of many substrates is effected at a high rate in membranes obtained from *G. liquefaciens* protoplasts (Stouthamer, 1961).

Inhibition of oxidation of some substrates and the complete absence of inhibition of oxidation of others are often observed. For instance, in particles obtained at 100,000 *g* from a *Pseudomonas sp.* cell homogenate, the oxidation of glucose was just as rapid as in cells, but oxidation of other sugars was much slower (Bentley and Schlechter, 1960). It has been suggested that the physical state of the cytoplasmic membrane controls respiratory activity in some way, as has been shown for *E. coli* (Henneman and Umbreit, 1964). It has been reported for *P. vulgaris* that destruction of the cells in a Hughes press greatly reduces the oxidation of lactate, succinate, and some other substrates of the tricarboxylic acid cycle (Jones and King, 1964).

Smith (1962) gives very interesting information on the effect of swelling and the concomitant deformation of the membrane on the operation of the respiratory chain. Spheroplasts of *B. subtilis* were obtained in 0.5 *M* sucrose and then put into solutions with a sucrose concentration of 0.2–0.8 *M*. In solutions with a sucrose concentration of 0.2–0.4 *M* the oxidation of endogenous substrates and amino acids was greatly inhibited, but the oxidation of NADH was increased. Smith suggests that some parts of the carrier membrane are not deformed at all and that the inhibition of oxidation of endogenous substrates and amino acids can be attributed to the detachment of weakly bound dehydrogenases and the failure of electrons to enter the chain. In fact, NADH oxidase is extremely stable, is usually found in the membranes, and has high activity. This stability of NADH oxidase can probably be attributed to the special strength of the bond of NADH dehydrogenase with the respiratory chain. Other dehydrogenases may be more weakly bound, so that deformation of the membrane leads to their detachment. In this way, the oxidation of several substrates is inhibited.

The succinate oxidation system in bacteria is particularly delicate (Repaske, 1954; Tucker, 1960). Lysis of *Corynebacterium erythrogenes* cells with lysozyme reduces the succinic oxidase activity by 90%. Another example of this is provided by *M. lysodeikticus* membranes. When protoplasts or even intact cells are lysed, the NADH oxidase in the membranes remains highly active, whereas succinic oxidase is greatly inhibited. The inhibition of succinic oxidase is probably due to the detachment of succinic dehydrogenase during lysis. The entire respiratory chain does not suffer in this case (Gel'man, Zhukova, Lukoyanova and Oparin, 1959; Gel'man, Lukoyanova, Zhukova, and Oparin, 1963).

Generally speaking, the connection of different dehydrogenases with the respiratory chain varies with the species of bacteria. For instance, it has been shown for *H. parainfluenzae* that formic, succinic, and NADH dehydrogenases are bound more tightly with the chain than D- and L-lactic dehydrogenases (White, 1965b). Lactate oxidation in ultrasonically disintegrated *C. diphtheriae* cells is effected at 10% of the rate in intact cells; succinic oxidase and NADH oxidase are not affected (Scholes and King, 1965).

The importance of the native structural state of the cytoplasmic membrane for the operation of the respiratory chain is clearly illustrated by the experiments of Gershanovich et al. (1963). Multiple injury to the cytoplasmic membrane of *E. coli* by "ghosts" of T-even phages inhibited respiration, NADH oxidation, and several other processes. The degree of inactivation was reduced by magnesium ions or the spermine polycation, probably because the deformation of the membrane was reduced (Gershanovich et al., 1966).

Another method of causing deformation of the membranes is to use chelating agents, which bind ions of bivalent metals, primarily magnesium. It has been noted in several investigations that the removal of magnesium ions from membranes or their fragments has an inhibiting effect on the respiratory chain. Glucose oxidation by particles obtained from *Pseudomonas* cell homogenates is completely suppressed by EDTA (Bentley and Schlechter, 1960). Similar effects have been observed for oxidation of various substrates by *A. aerogenes* particles (Blakley and Cifferi, 1961).

A chelating agent of the EDTA type can presumably affect the function of the respiratory chain by its ability to destroy the ultrastructure of the membrane. *P. striata* particles (100,000 *g*) containing D-allohydroxyproline oxidase are dissolved by 0.025 *M* EDTA. This, of course, leads to inactivation of the oxidase. The respiratory chain is broken up into fragments, one of which, D-allohydroxyproline dehydrogenase, has been isolated (Yoneya and Adams, 1961; Adams and Newberry, 1961).

Inhibition of the respiratory chain of *M. lysodeikticus* by EDTA and the concomitant destruction of the membranes has been reported (Lukoyanova *et al.,* 1961). In this case an enzyme–lipid complex containing malic dehydrogenase, NADH dehydrogenase, and lipids passed into solution (Gel'man, Zhukova and Oparin, 1963).

The ability of EDTA to destroy membranes and liberate enzyme–lipid complexes or even individual enzymes has also been used in work with mitochondria. For instance, cytochrome oxidase can be extracted from beef heart or yeast mitochondria with 0.05 *M* Na$_4$ EDTA and 0.05 *M* Tris at pH 9.0 (Person and Zipper, 1965).

Magnesium ions have been shown to have a stabilizing effect on bacterial membranes (Weibull, 1956; Storck and Wachsman, 1957; Zickler, 1965). The behavior of membranes obtained by osmotic shock from *Pseudomonas* protoplasts depended on the presence of magnesium ions (5 × 10^{-3} *M*). Membranes obtained in the absence of magnesium do not oxidize succinate and malate at all, whereas those obtained in a medium containing Mg^{++} oxidized these substrates well if coenzymes were added to the medium (Mizuno *et al.,* 1961). It is reported that oxidation in *M. lysodeikticus* membranes can be stabilized by magnesium ions, and also by spermine and spermidine (Ishikawa and Lehninger, 1962). Yet, inhibition of the respiratory chain by EDTA cannot be reversed by magnesium ions, as has been shown for *Pseudomonas fragii* particles in the case of oxidation of arabinose, xylose, and ribose (Weimberg, 1963).

The effects of EDTA, magnesium ions, and spermine on the respiratory chain can probably be attributed to a particular charge distribution on the membrane, the liberation or protection of hydrophilic groups, increased solubilization, or, on the contrary, strengthening of the bond of the enzymes with the respiratory chain system. There are cases, however, where stabilization of the respiratory chain is effected by a more obscure mechanism. An investigation of the operation of the respiratory chain in *M. lysodeikticus* membranes showed that membranes obtained without deoxyribonuclease treatment of the bacterial lysate oxidized malate and succinate better than those obtained after degradation of the DNA (Gel'man, Zhukova and Oparin, 1959; Lukoyanova *et al.*, 1963). The protective effect of DNA was manifested only when the magnesium ion concentration was high (2×10^{-2} M).

SPECIAL FEATURES OF THE BACTERIAL RESPIRATORY CHAIN

An examination of bacterial respiratory chains, their assembly, functioning, and response to inhibitors reveals several special features.

Effect of Inhibitors

Inhibitors of mitochondrial respiratory chains do not always have the same effect on the electron-transport chains of bacteria. The results of experiments investigating the effect of inhibitors on intact bacteria are not completely reliable, since the inhibitor may have difficulty penetrating into the cell. Hence, it is better to work with disrupted cells.

The generally recognized inhibitors of cytochrome components of the respiratory chain are cyanide, azide, and carbon monoxide. The difficulties of work with cyanide and azide have been discussed in the literature (L. Smith, 1954*a*, 1963; Mikhlin, 1960). A more reliable test for group *A* cytochromes in bacteria consists in suppression of respiration by carbon monoxide and the spectral investigation of the complex formed between the cytochrome and CO. All the investigated species of bacteria containing group *A* cytochrome form this complex. Cytochrome *o* can also form it (Castor and Chance, 1959; Rees and Nason, 1965; Gray *et al.*, 1966).

Respiration of cell-free extracts of many bacteria is suppressed by 10^{-3} M cyanide (Smith, 1954*a*, *b*, *c*; Asnis *et al.*, 1965; Keilin and Harpley, 1941; Mathews and Sistrom, 1959). Cyanide inhibits cytochromes $a_1 + a_2$ of *E. coli* and also inhibits (by 50%) cytochromes $a + a_3$ of a fast-growing strain of the diphtheria bacterium (Kashket and Brodie, 1963a; Pappenheimer *et al.*, 1962). Cyanide inhibition is also found in *A. vinelandii* (Layne and Nason, 1958; Repaske and Josten, 1958). The cytochrome system ($a + a_3$) of vegetative cells of *M. xanthus* is 60% inhibited by cyanide (Dworkin and Niederpruem, 1964). *Pseudomonas sp.* and *H. parainfluenzae* have a cyanide-sensitive respiratory chain (Francis *et al.*, 1963; White and Smith, 1964). The degree of inhibition in preparations of different bacteria produced by cyanide in one concentration varies (Table 19).

Cytochrome *a*, as Keilin originally suggested, does not interact with cyanide. This applies both to mitochondrial cytochrome *a* and to the bacterial forms (Keilin and Hartree, 1938, 1939; Horie and Morrison, 1963; King and Kuboyama, 1964; Lemberg *et al.*, 1964; Slater, 1964).

It can be assumed that cyanide resistance to cytochrome-containing respiratory chains means that the bacteria have cytochrome *a* (α-band 601 mμ) as a terminal oxidase. This may account for the resistance of NADH oxidase of preparations from *M. tuberculosis* and *M. avium* to cyanide (5×10^{-3} M) (Goldman *et al.*, 1963; Kusunose and Goldman, 1963). Cyanide also fails to inhibit oxidation in preparations from *M. lysodeikticus* (Gel'man, Zhukova, Lukoyanova and Oparin, 1959; Mitchell, 1962; Ishikawa and Lehninger, 1962; Gel'man *et al.*, 1963). According to Japanese authors, however, 10^{-3} M potassium cyanide causes 50% inhibition of NADH oxidase (Fujita *et al.*, 1966). Azide in a concentration of 10^{-2} M does not inhibit the respiratory chain of this species (Mitchell, 1962).

Azide is often a less specific inhibitor of bacterial cytochrome oxidase (Bruemmer *et al.*, 1957; Tissières, 1951; Vernon and White, 1957). White showed that cyanide and carbon monoxide inhibited the respiratory chain of the facultative anaerobe *H. parainfluenzae*, whereas azide had no effect (White and Smith, 1962). Rotenone (fish poison) in very low concentration (25 mμmole per gram of protein) is a powerful inhibitor of mitochondrial NADH dehydrogenase (Burgos and Redfearn, 1965), but does not inhibit succinic oxidase. Rotenone does not inhibit NADH oxidase or lactic oxidase in cell-free preparations of *H. parainfluenzae* (White and Smith, 1964). It has been reported in several cases that agents which uncouple oxidative

Table 19. Effect of Cyanide on Bacterial Respiratory Chain

Bacterium	Conc., M	Enzyme system	Inhibition, %	Author
Pseudomonas sp.	10^{-3}	Malic oxidase	50–80	Francis *et al.*, 1963
Pseudomonas aeruginosa	$3 \cdot 10^{-4}$	Cytochrome c_{551}:nitrite:O_2 oxidoreductase	96	Yamanaka and Okunuki, 1963*b*
Pseudomonas aeruginosa	10^{-4}	Cytochrome c:O_2 oxidoreductase	96	Azoulay and Couchoud-Beaumont, 1965
Aerobacter aerogenes	10^{-3}	Succinic oxidase	95	Tissières, 1951
	10^{-3}	Pyruvic oxidase	95	
	10^{-4}	Pyruvic oxidase	50	
Azotobacter vinelandii	10^{-3}	NADH oxidase	85	Repaske and
	$2 \cdot 10^{-4}$	NADH oxidase	48	Josten, 1958
Azotobacter vinelandii	10^{-2}	Succinic oxidase	80	Repaske, 1954
Azotobacter vinelandii	10^{-3}	NADH:nitrate reductase	70–90	Taniguchi and Ohmachi, 1960
Azotobacter vinelandii	10^{-3}	NADH oxidase	100	Jones and
		Succinic oxidase	100	Redfearn, 1966
Pasteurella tularensis	$6 \cdot 10^{-2}$	NADH oxidase	80	Robinson and Mills, 1961
Staphylococcus aureus	10^{-3}	NADH oxidase	60	Taber and Morrison, 1964
Mycobacterium phlei	$3 \cdot 10^{-3}$	Succinic oxidase	95	Asano and
		NADH oxidase	55	Brodie, 1964
Acetobacter xylinum	10^{-3}	Malic oxidase	100	Benziman and Perez, 1965
Sarcina lutea	10^{-3}	Succinic oxidase	96	Mathews and Sistrom, 1959
Escherichia coli	$7 \cdot 10^{-3}$	Succinic oxidase	100	Kashket and
		Malic oxidase	100	Brodie, 1963
Corynebacterium diphtheriae PW8$_8$P	$2.5 \cdot 10^{-3}$	Succinic oxidase	50–60	Pappenheimer *et al.*, 1962
Myxococcus xanthus cells	10^{-3}	NADH oxidase	78	Dworkin and Nieder-
Myxococcus xanthus microcysts	10^{-3}	NADH oxidase	29	pruem, 1964
Micrococcus denitrificans	10^{-4}	Cytochrome c oxidase	100	Vernon, 1957
Acetobacter peroxydans	10^{-3}	Lactic oxidase	90	De Ley and Schell, 1959
Xanthomonas phaseoli	$5 \cdot 10^{-3}$	Succinic oxidase	87	Madsen, 1960
Proteus vulgaris	$3.3 \cdot 10^{-3}$	NADH oxidase	84	Feldman and O'Kane, 1960

phosphorylation in mitochondria (2,4-dinitrophenol and Dicumarol) inhibit the respiration of cell-free bacterial preparations (White and Smith, 1964).

Antimycin A is a strong inhibitor of the mitochondrial respiratory chain; its site of action lies somewhere between cytochromes b and c (Chance and Williams, 1956). There are several views on the mechanism of action of this inhibitor in the mitochondrial respiratory chain. According to Green (1962), antimycin A reacts in the mitochondrial respiratory chain with the carrier which follows cytochrome b, probably nonheme iron. It has been suggested that antimycin A prevents the breakdown of the complex of cytochrome b and c_1 by acting on the chemical grouping responsible for the binding of these cytochromes (Rieske and Zaugg, 1962). Antimycin A may act on the lipid factor located between cytochrome b and c (Sazykin, 1965). Despite these various opinions, however, the chemical nature of the antimycin-sensitive factor in mitochondria remains essentially unknown (Estabrook, 1962).

The literature devoted to bacterial respiratory systems includes only a few cases of inhibition by antimycin A, and then at concentrations 10^6 times greater than those which affect mitochondria (10^{-9} M in mitochondria, up to 10^{-3} M in bacteria). Straat and Nason (1965) showed that 2×10^3 M antimycin A causes 90% inhibition of the nitrite reductase system in $N.$ $agilis$. Similar high concentrations of antimycin A inhibit nitrite oxidase by 60% in $N.$ $winogradskii$ (Van Gool and Laudelout, 1966). The succinic oxidase of $A.$ $vinelandii$ is 18% inhibited by antimycin A in a concentration of 5 µg/8µg of protein (Repaske, 1954). It was believed at first that low sensitivity to antimycin A was due to the failure of the lipid-soluble antimycin A to penetrate the bacterial cell (Ahmad et $al.,$ 1950; Smith, 1954$a,$ b). Experiments were conducted with cell-free extracts of bacteria: $P.$ $fluorescens,$ $E.$ $coli,$ $S.$ $aureus,$ $P.$ $vulgaris,$ $B.$ $subtilis,$ $M.$ $lysodeikticus,$ $A.$ $vinelandii,$ $C.$ $diphtheriae,$ etc. These experiments again showed the absence of an effect due to antimycin A (Goldman et $al.,$ 1963; Pappenheimer et $al.,$ 1962; Slater, 1961; Lightbown and Jackson, 1956; White, 1963; Taber and Morrison, 1964; Jackson and Lawton, 1958; Ramachandran and Gottlieb, 1961; Gel'man et $al.,$ 1963; Scholes and King, 1965a). It appears that bacterial respiratory chains lack the antimycin-sensitive factor present in mitochondria. The interpretation of the information relating to the inhibiting effect of antimycin

A on bacteria is complicated by the fact that, unlike the situation in mitochondria, antimycin A causes destruction of the cytoplasmic membrane, a change in permeability, and cell death (Marquis, 1965).

Some derivatives of quinoline N-oxide act on bacteria in the same way as antimycin A acts on mitochondria. It has been suggested that inhibition of bacterial respiratory chains by 2-*n*-heptyl 4-hydroxyquinoline N-oxide (NOQNO) is because its molecule is smaller than that of antimycin A (Sazykin, 1965). This substance inhibits the respiratory chains of several species of bacteria, including *H. parainfluenzae, E. coli, S. aureus,* and *A. vinelandii* (White and Smith, 1962, 1964; Kashket and Brodie, 1963*a, b*; Taber and Morrison, 1964; Jones and Redfearn, 1966).

It is only in bacteria that we find varying reactions to inhibitors. This is due to the induced synthesis of particular respiratory systems. For instance, the fast-growing strain of *C. diphtheriae* contains a complete set of cytochromes: *a*, a_3, *b*, and *c*. Its respiration is inhibited by carbon monoxide and cyanide. The toxic strain of this species contains only cytochrome b_{558}, which effects succinate oxidation. The succinic oxidase of this strain, of course, cannot be inhibited by cyanide or carbon monoxide (Pappenheimer *et al.,* 1962). Considerable variations in the composition and concentration of respiratory chain components have been reported for a hemin-requiring strain of *Hemophilus sp.* This leads to different responses to inhibitors (White, 1963).

Azoulay (1964) investigated the induced synthesis of two different terminal oxidases in *P. aeruginosa.* The cytochrome c_{551}:nitrite: O_2 oxidoreductase was sensitive to 3×10^{-4} *M* cyanide. The cytochrome *c*:O_2 oxidoreductase (EC 1.9.3.1) was inhibited only by 10^{-3} *M* potassium cyanide.

Differences in cyanide inhibition in relation to phrases of growth were observed in heterotrophically grown *R. rubrum* (Taniguchi and Kamen, 1965). In the exponential phase 10^{-3} *M* KCN completely inactivated succinic oxidase and NADH oxidase. A concentration of 2×10^{-4} *M* led to 50% inhibition in this phase. In the stationary phase, a concentration three times greater (6×10^{-4} *M*) was required for 50% inhibition. The authors think that the relative importance of the terminal oxidases (cytochrome *o* and the autoxidable cytochrome *b*) is different at different phases of growth.

It has been shown that oxygen tension determines the amount of cytochromes formed and the nature of the terminal electron transport chain in *Achromobacter* (Arima and Oka, 1965*a*). In this case cytochrome a_2 is relatively insensitive to cyanide and plays the major role in cyanide-resistant respiration; it is also the terminal oxidase in anaerobic conditions.

Very little is known regarding the effect of inhibitors on the initial, noncytochrome portion of the bacterial respiratory chain, which includes dehydrogenases of flavoprotein nature. Inhibitors of flavin enzymes include quinine, atebrin, amobarbital, and other derivatives of barbituric acid, but the mechanism of their action and their specificity have not been conclusively established (Beinert, 1960; Singer, 1963). Amytal in concentration 6.6×10^{-3} M inhibits the succinic oxidase of *Xanthomonas phaseoli* by 65% (Madsen, 1960), 10^{-3} amobarbital inhibits NADH reductase by 55% (Taniguchi and Ohmachi, 1960), and 2×10^{-3} M amobarbital strongly inhibits (by up to 80%) malic oxidase from *Pseudomonas sp.* and *A. xylinum* (Francis *et al.*, 1963; Benziman and Perez, 1965). Yet 5×10^{-3} M amobarbital has no effect at all on lactic oxidase from *A. peroxydans* or NADH oxidase from *S. aureus* (De Ley and Schell, 1959; Taber and Morrison, 1964). White (1964) showed that secobarbital acts on a respiratory-chain dehydrogenase. In particles from heterotrophically grown *R. rubrum*, the respiratory chain is completely inhibited by amobarbital (Taniguchi and Kamen, 1965). Amobarbital (10^{-2} M) also has a slight effect on the respiratory chain of *M. phlei;* the inhibition of NADH oxidase does not exceed 35% (Asano and Brodie, 1964).

There is rather contradictory information on the effect of atebrin and quinine on the respiratory chain. Atebrin (1×10^{-3} M) does not inactivate the lactic oxidase of *A. peroxydans* (De Ley and Schell, 1959), but inhibits NADH oxidase in preparations from *Pasteurella tularensis* (Robinson and Mills, 1961). Atebrin (10^{-3} M) inhibits nitrite oxidase by only 25% in *Nitrobacter* particles (Aleem and Nason, 1959), yet 3×10^{-3} M atebrin inhibits succinic oxidase by 85% in *M. phlei* particles (Asano and Brodie, 1964).

Quinone (10^{-3} M) inhibited NADH:nitrate reductase by 60% (Taniguchi and Ohmachi, 1960) and nitrite oxidase by only 30% (Aleem and Nason, 1959) in *Nitrobacter* particles. It has been shown that some isolated dehydrogenases are inactivated by amobarbital in the presence of artificial hydrogen acceptors. For instance, the malic dehydrogenases of *M. avium* and *M. phlei,* the prosthetic group of which apparently consists of FAD and FMN, are greatly inhibited by

10^{-3} amobarbital (Kimura and Tobari, 1963; Asano and Brodie, 1965a). Amobarbital also inhibits the flavin malic dehydrogenase isolated from *Acetobacter xylinum* (Benziman and Galanter, 1964). Luminal (8×10^{-3} M) inhibits the NADH dehydrogenase of *M. lysodeikticus* by 55% (Gel'man, Zhukova and Oparin, 1963).

A special case in bacteria is the replacement of cytochrome respiratory chains by flavin respiratory chains, thus altering the response to inhibitors. This has been observed in the vegetative cells and spores of *B. cereus* and in the microcysts and vegetative cells of *M. xanthus* (Doi and Halvorson, 1961; Dworkin and Niederpruem, 1964). The authors note that NADH oxidase in spores and cysts is less sensitive to cyanide, which has an inhibiting effect on cellular NADH oxidase. The oxidation of NADH in spore preparations is sensitive to atebrin, an inhibitor of flavin enzymes (Doi and Halvorson, 1961).

It is clear from this short review of the effect of inhibitors on bacterial electron-transport chains that the recognized inhibitors of mitochondrial electron transport often do not have the effect one would expect on bacterial respiratory chains. The reasons for this require special investigation in each particular case.

Schemes of Bacterial Respiratory Chains

The lack of uniformity in the effects of inhibitors on the respiratory chain in bacteria apparently indicates differences in enzyme composition. It was shown above that the enzyme composition of the respiratory chain in different bacteria may vary as regards set of cytochromes, dehydrogenases, and quinones, but these differences are particularly distinctly revealed by a comparison of the enzyme composition and the sequence of respiratory chain components as a whole. As an example we give the respiratory chains of some of the most thoroughly investigated bacteria. The respiratory chain of *M. phlei* is probably most like that of mitochondria. Cytochromes $a + a_3$ are present, and HCN and CO cause inhibition. The set of other cytochromes (b, c_1, c) is also very similar to that of mitochondria. The succinic dehydrogenase and NADH dehydrogenase behave like typical flavin enzymes and are inactivated by atebrin. Yet ubiquinone is replaced by naphthoquinone and *M. phlei* has a special enzyme (malate:vitamin K reductase) which introduces an electron from malate into the chain.

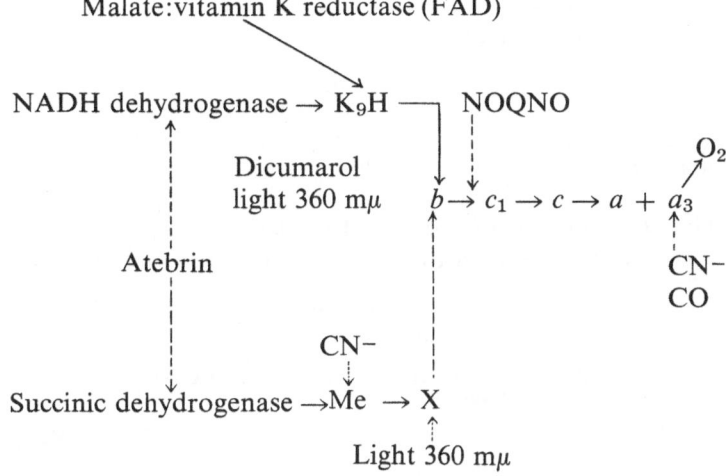

Respiratory chain of *Mycobacterium phlei* (Asano and Brodie, 1964, 1965*b*). Light 360 mμ means inhibition on illumination; NOQNO is 2-*n*-heptyl-4-hydroxyquinoline N-oxide; K$_9$H is vitamin K$_{2(45)}$; Me here and henceforth is metal; X here and henceforth is an unidentified component.

The respiratory chain of *C. diphtheriae* is similar in basic features to that of *M. phlei,* but the enzyme composition depends considerably on the strain. For instance, cytochrome b_{558}, which interacts with oxygen, predominates in the slow-growing toxic strain.

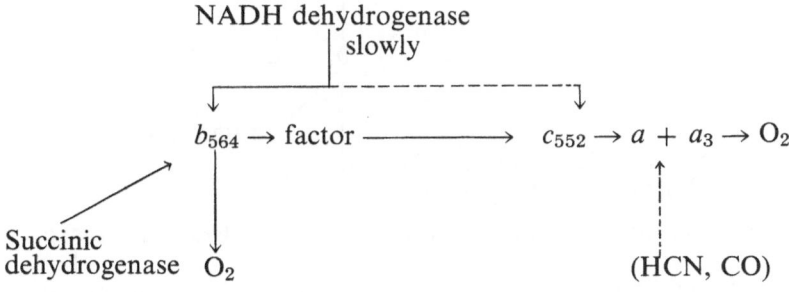

Respiratory chain of *Corynebacterium diphtheriae* (fast-growing wild strain) (Pappenheimer *et al.,* 1962).

Succinic
dehydrogenase → b_{558} → O_2

$(a + a_3 → O_2)$

Respiratory chain of slow-growing strain PW8(Pd) of *Corynebacterium diphtheriae* (Pappenheimer *et al.*, 1962; Kleczkowska *et al.*, 1965).

The electron-transport chain of *C. diphtheriae* strain CN 2000, the producer of diphtheria toxin, has been thoroughly investigated.

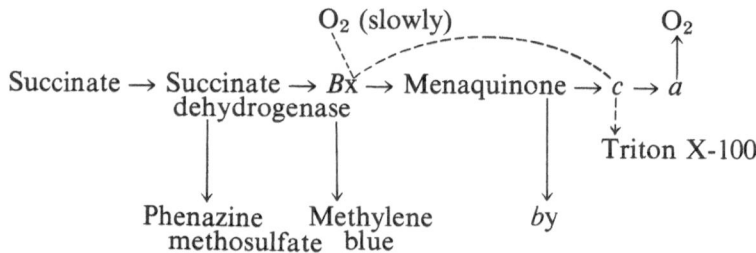

Sucinnic oxidase of *Corynebacterium diphtheriae* (Scholes and King, 1965).

In most bacteria the respiratory chains differ even more from the mitochondrial respiratory chain in their set of components. *E. coli* has alternative pathways of electron transport to cytochrome *b* from NADH and succinate. In the first case, the transfer proceeds via vitamin $K_{2(45)}$, and in the second, via UQ_8. Formic dehydrogenase and nitrate reductase are linked to the chain.

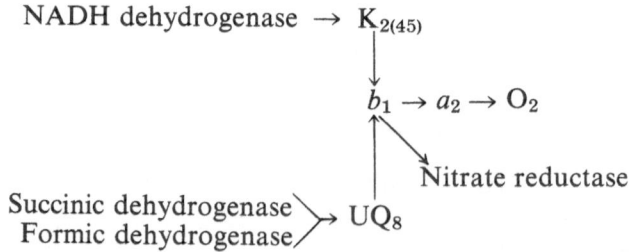

Respiratory chain of *Escherichia coli* (Kashket and Brodie, 1963*b*; Itagaki, 1964).

Despite the numerous investigations of respiration in *Azotobacter,* the compilation of a single scheme for the respiratory chain is difficult. As a tentative attempt we give the scheme below. As in other bacteria, there is a large set of group *A* cytochromes, the functions of which have not yet been distinguished. Cytochromes c_4 and c_5 are found only in *Azotobacter*.

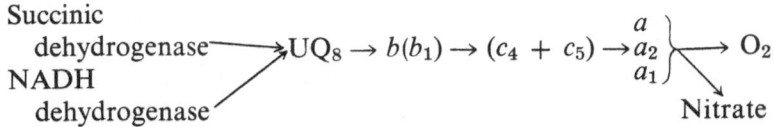

Respiratory chain of *Azotobacter vinelandii* (Tissières, 1956; Tissières *et al.*, 1957; Bruemmer *et al.*, 1957; Hovenkamp, 1959a, b; Taniguchi and Ohmachi, 1960; Temperli and Wilson, 1962; Jones and Redfearn, 1966).

Distinctive respiratory chains are found in acetic acid bacteria, which have no cytochromes $a + a_3$ and are rich in tightly bound dehydrogenases.

D-Lactic dehydrogenase

L-Lactic dehydrogenase $\longrightarrow c_{555} \rightarrow c_1 \rightarrow a_1(a_2) \rightarrow O_2$

Alcohol dehydrogenase

Respiratory chain of *Acetobacter* (Iwasaki, 1960; De Ley and Schell, 1959; De Ley, 1960).

A special feature of the respiratory chain of several *Pseudomonas* species is the presence of the noncytochrome blue protein in the chain. The terminal oxidases—cytochrome oxidase and nitrite reductase—differ in chemical nature and are inhibited by cyanide in various concentrations. Ubiquinone has been found in *P. aeruginosa* (Bishop *et al.*, 1962), but its role in the respiratory chain has not been investigated.

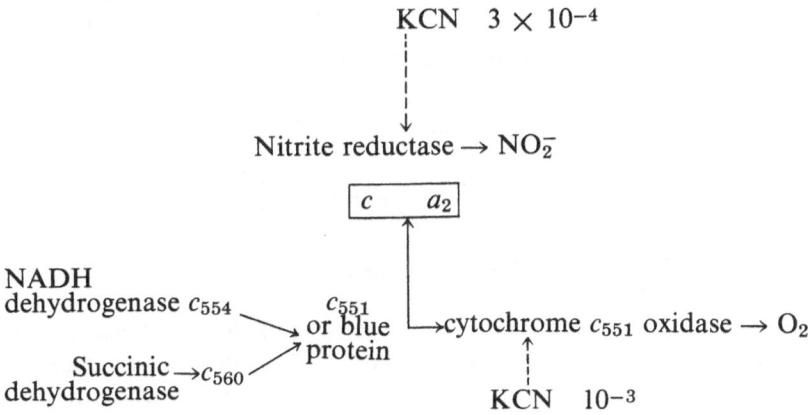

Respiratory chain of *Pseudomonas aeruginosa* (Yamanaka, 1964; Horio *et al.*, 1961a, b; Azoulay, 1964; Azoulay and Couchoud-Beaumont, 1965).

The respiratory chain of a halotolerant micrococcus contains an antimycin-sensitive link and a noncytochrome brown protein, which is implicated in nitrate and oxygen respiration, as is the case for the blue protein.

$$FP \rightarrow b_{560} \rightarrow X \begin{array}{l} \nearrow c_{554}\ (\text{I}) \rightarrow \text{Brown protein} \rightarrow \text{Nitrate} \\ \searrow c_{554}\ (\text{II}) \rightarrow c_{551} \rightarrow o \end{array}$$

$$\uparrow$$
$$\text{Antimycin A} \qquad\qquad O_2 \downarrow$$

Respiratory chain of *Micrococcus sp.* (Hori, 1963).

The respiratory chain of cyanide-resistant *Achromobacter* cells can contain three terminal oxidases—cytochromes o, a_1, a_2. According to the authors, cytochrome a_2 is a cyanide-resistant oxidase.

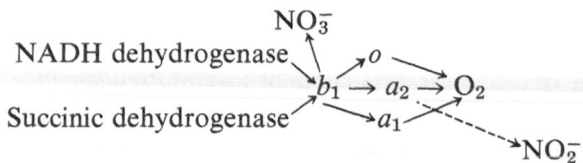

Respiratory chain of cyanide-resistant *Achromobacter* cells (Arima and Oka, 1965*a*).

The respiratory chain of *H. parainfluenzae* is also characterized by firmly bound dehydrogenases. Vitamin K is replaced by demethyl vitamin K_2. The terminal oxidase is cytochrome a_2; cytochrome o or a_1 may function along with cytochrome a_2.

$$FP \rightarrow X \rightarrow \text{Demethyl vitamin } K_2 \rightarrow b_1 \rightarrow c_1 \rightarrow a_2 \rightarrow O_2$$

$$\uparrow \qquad\qquad\qquad\qquad\qquad\qquad\qquad \uparrow$$
$$\text{Secobarbital} \qquad\qquad \text{NOQNO} \qquad\qquad \text{KCN}$$

Respiratory chain of *Hemophilus parainfluenzae* (White, 1965*a*).

We suggest that the terminal oxidase of *M. lysodeikticus* is cytochrome *a*. The respiratory chain contains a firmly bound malic dehydrogenase. It has been suggested that the terminal oxidase may be cytochrome *b*.

NADH dehydrogenase $\searrow$
 Malic dehydrogenase $\rightarrow$ [Vitamin $K_{2(45)}$] $\rightarrow b_{560} \rightarrow c_{550} \rightarrow a_{601}$
Succinic dehydrogenase $\nearrow$ $\qquad$? $\qquad\qquad$ $\downarrow$ $\qquad\quad$ $\downarrow$
$\qquad\qquad\qquad\qquad\qquad\qquad\qquad\qquad\qquad$ O_2 $\qquad\quad$ O_2

Respiratory chain of *Micrococcus lysodeikticus* (Gel'man, Lukoyanova, Zhukova, and Oparin, 1963; Bishop and King, 1962; Ishikawa and Lehninger, 1962; Mitchell, 1962; Fujita *et al.*, 1966).

The respiratory chain of a pathogen which produces gall tumors in plants contains flavoprotein, UQ_{10}, and a type *c* cytochrome with absorption maxima at 550, 520, and 420 mμ.

NADH $\longrightarrow$ Flavoprotein $\rightarrow UQ_{10} \rightarrow$ Cytochrome $\longrightarrow O_2$

 Amobarbital $\qquad\qquad\qquad$ $\diagdown$ $\diagup$ $\qquad\qquad$ Cyanide

$\qquad\qquad\qquad\qquad\qquad$ Menadiol

Respiratory chain of *Agrobacterium tumefaciens* (Kurup *et al.*, 1966).

The respiratory chains of chemoautotrophs have not received much study. Some of the information has been reviewed (Doman and Tikhonova, 1965). In addition to a fairly typical respiratory chain each of the species contains supplementary dehydrogenases which transfer electrons from the inorganic substrate to the chain.

H_2
$\qquad\qquad$ $\searrow$
$\qquad\qquad\quad \nearrow$ NAD $\rightarrow$ FMN $\rightarrow b \rightarrow c \rightarrow$ Cytochrome oxidase
$\qquad \nearrow$ $\qquad\qquad\qquad\qquad\qquad\qquad\qquad$ $\diagup$ $\diagdown$
Lactate $\qquad\qquad\qquad\qquad\qquad\qquad\qquad$ O_2 $\quad$ Nitrate

Respiratory chain of *Hydrogenomonas* (Packer, 1958*a*, *b*; Lees, 1958).

The same principle of organization can be seen in the chain of *Nitrosomonas* and *Nitrosocystis*.

$$NH_2OH \rightarrow FMN \rightarrow b \rightarrow c \rightarrow a_1 \rightarrow a$$
$$\downarrow$$
$$O_2$$

Respiratory chain of *Nitrosomonas* and *Nitrosocystis* (Aleem and Lees, 1963; Falcone *et al.*, 1962, 1963; Hooper and Nason, 1965).

The respiratory chain of *Nitrobacter* does not have type *B* cytochrome, but apparently contains several group *A* cytochromes.

$$NO_2^- \xrightarrow{} c \begin{array}{c} \nearrow a_{1(583)} \to (?) \\ \\ \longrightarrow a_{1(587)} \to \end{array} \begin{array}{c} \nearrow Me\ (?) \to Nitrate \\ \\ \searrow a_{(605)} \longrightarrow O_2 \end{array}$$

KCN
KCIO$_3$

KCN
NaN$_3$

CO

Respiratory chain of *Nitrobacter* (Aleem and Nason, 1960; Straat and Nason, 1965).

In the respiratory chain of iron bacteria an electron from Fe++ also arrives at cytochrome *c*. No group *B* cytochromes can be detected (Vernon *et al.,* 1960; Dugan and Lundgren, 1965).

The special features of the respiratory chains of photoheterotrophs have already been discussed.

Differences in the composition of the respiratory chains in separate representatives of the bacteria relate to the composition of the terminal oxidases, group *B* and *C* cytochromes, ubiquinones, and naphthoquinones. Group *A* cytochromes are not the terminal oxidases in all bacteria. For instance, in *C. diphtheriae* and *R. rubrum,* oxygen is reduced by group *B* cytochromes. The group *B* and *C* cytochromes also vary greatly. In most bacteria, the actimycin-sensitive factor is absent. The composition of the dehydrogenases beginning the respiratory chain also varies. Succinic and NADH dehydrogenases are the most constant components, but, in addition to them, the enzyme system may contain malic dehydrogenase, formic dehydrogenase, lactic dehydrogenase, alcohol dehydrogenase, etc. Thus, even this cursory review indicates that the composition of the respiratory chain in different bacteria is not the same, much less in bacteria and mitochondria.

Functional Regulation of Respiratory Chain

The special nature of the bacterial respiratory apparatus is also evident in other factors: the structure of the membrane systems, the relationship between the respiratory chain and the cytoplasm, and the mode of regulation.

As already mentioned, the respiratory chain in the bacterial cell can be mounted on the cytoplasmic membrane and membranous structures. These structures are basically unit membranes, and it is only in some bacteria that they are "packed" in a complex manner. For an understanding of the special features of functioning of the bacterial respiratory apparatus it is important to recall the absence of a limiting envelope in the membranous structures. In bacteria, the "open" membrane system is in direct contact with the cytoplasm—immersed in it and not separated by a special barrier such as the external limiting membrane of mitochondria. This, of course, applies also to bacteria in which the membranous structures are poorly developed, and the respiratory chain is localized mainly in the cytoplasmic membrane. Hence, in bacteria there is a direct interaction between the membranes carrying the respiratory chain and the cytoplasm, which is the source of coenzymes, substrates, and ADP. Since the bacterial cell contains no mitochondria, the metabolites which accumulate in the cytoplasm as a result of anaerobic degradation of substances (pyruvate, lactate, ethanol) are rapidly oxidized by the enzymes in the membranes (White, 1966).

Special note should be taken of the ease of oxidation of NADH, the universal intermediate in the transfer of hydrogen between anaerobic and aerobic enzyme systems. The membrane systems of bacteria very rapidly oxidize NADH, the oxidation of which in mitochondria is always impeded by the low permeability of the outer mitochondrial envelope to this coenzyme.

Membranes extracted from bacteria contain hardly any P_{inorg} or NAD (Ostrovskii and Gel'man, 1963; Gel'man, Zhukova and Zaitseva, 1962; Jones and Redfearn, 1966). These low-molecular components are removed during extraction and washing of the material. Mitochondria always retain a certain amount of these compounds. Here, then, is another difference between bacterial membranes and mitochondria due to the absence of formed organelles in bacteria.

Owing to the small size of the bacterial cell, metabolites and respiratory apparatus are close together. This is probably why the respiratory apparatus of bacteria contains dehydrogenases capable of oxidizing metabolites without the aid of coenzymes, which are intermediates between the metabolites and the respiratory chain in most cells. Acetic acid bacteria provide an example of the interaction of the membranous respiratory chain with the enzyme systems of the cytoplasm. Particles from *A. peroxydans* contain the complete system of enzymes necessary to oxidize D- and L-lactate, pyruvate, ethanol, and acetaldehyde to acetate. The dehydrogenating enzymes interact with the respiratory chain localized in these particles. Thus, several oxidative transformations occur in the membrane system, and the substrate enters from the cytoplasm (De Ley and Schell, 1959). From an investigation of acetic acid bacteria, De Ley and Schell 1962 derived the following scheme to illustrate the interaction of the membrane system and the cytoplasm in the oxidation of substrates.

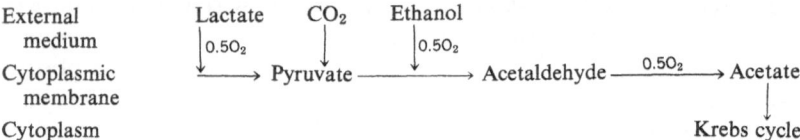

Unfortunately, very few bacteria have been investigated from the viewpoint of the distribution of oxidation–reduction enzymes between the cytoplasm and the membrane system. As an example we note that in *P. fluorescens* cells the enzymes of the pentose phosphate cycle—phosphohexoisomerase, glyceraldehyde-3-phosphate dehydrogenase, glucose-6-phosphate dehydrogenase, and transketolase—are not connected with the membrane but are dissolved in the cytoplasm and, thus, intermediate products and NADH can easily diffuse to the membranes carrying the respiratory chain and be oxidized there (Burrous and Wood, 1962).

The lack of an envelope confining the membranous structures of bacteria and the possibility of free interaction between the metabolites and coenzymes of the cytoplasm with the respiratory chain give rise to some differences in the regulatory mechanisms of the bacterial respiratory chain. Bacterial cells do not have any mechanism regulating oxidative metabolism as a whole which would correspond to respiratory control. In the bacterial cell any available substrate is apparently oxidized rapidly, and such processes often proceed without any correlation with the energy requirements of the cell. The absence of mitochondria and the imperfect organization of the membranes in bacteria are of particular significance in the process of oxidative phosphorylation. This will be discussed in the next chapter.

Reverse electron transfer in the form established for mitochondria (Chance and Hagihara, 1962; Klingenberg and Schollmeyer, 1962) has not been found in bacteria. It is true that the energetics of some chemoautotrophs (*Nitrobacter, T. ferrooxidans*) suggests, even purely theoretically, the existence of an "energy lift," since the assimilation of CO_2 requires a reductant of the pyridine nucleotide type with a potential of -0.3 v, and the substrate for oxidation must have a higher potential (Lees, 1960; Zavarzin, 1964; Doman and Tikhonova, 1965). Such a process has recently been discovered in *N. agilis, N. europea, Thiobacillus novellus* and *T. ferrooxidans* (Aleem, 1966*a, b*; Tikhonova *et al.,* 1967). It was shown that cell-free preparations of *N. europea* catalyzed the reversal of electron transfer from ferrocytochrome *c* to NAD or NAD(P) with the consumption of ATP. The electron donors were hydroxylamine or succinate. Uncouplers of oxidative phosphorylation inhibited the reverse transfer. The following mechanism of the electron-transfer reactions in *N. europea* was suggested:

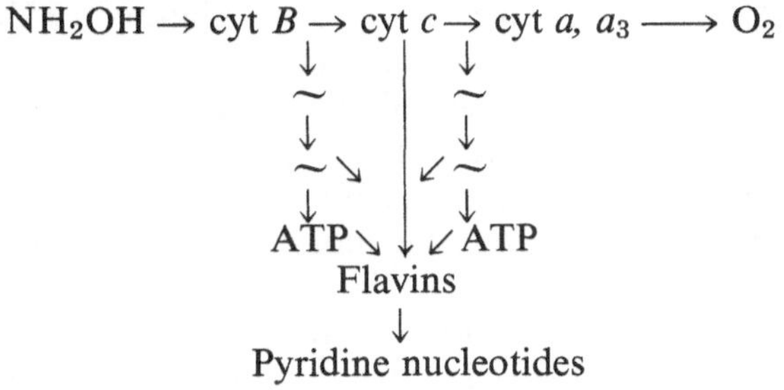

$$NH_2OH \rightarrow cyt\ B \rightarrow cyt\ c \rightarrow cyt\ a,\ a_3 \longrightarrow O_2$$

Flavins

Pyridine nucleotides

Cell-free extracts of *T. novellus* have also catalyzed ATP-dependent reduction of pyridine nucleotides (Aleem, 1966*b*). The electron donors were thiosulfate, formate, or mitochondrial reduced cytochrome *c*. Uncouplers of oxidative phosphorylation, as well as atebrin and amobarbital, inhibited the process. It is important to note that reverse transfer is found only in cells grown on an inorganic substrate, i.e., it is to some extent an adaptive process. The following scheme of electron transport for autotrophically grown *T. novellus* cells has been suggested:

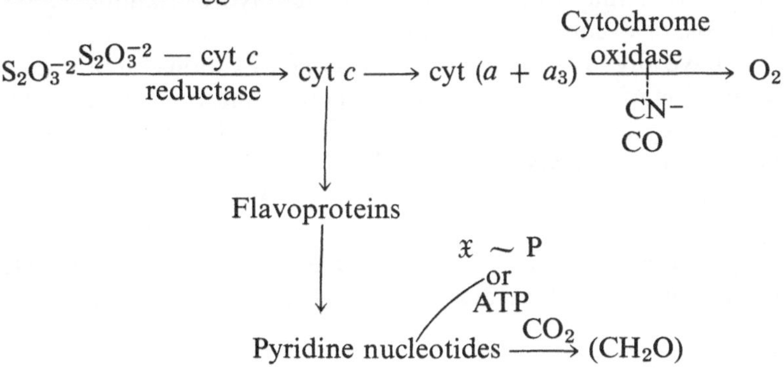

$$S_2O_3^{-2} \xrightarrow[\text{reductase}]{S_2O_3^{-2}\ -\ cyt\ c} cyt\ c \longrightarrow cyt\ (a + a_3) \xrightarrow[\substack{CN^- \\ CO}]{\text{Cytochrome oxidase}} O_2$$

Flavoproteins

$\mathfrak{x} \sim P$
or
ATP

Pyridine nucleotides $\xrightarrow{CO_2}$ (CH_2O)

Cell-free extracts of *T. ferrooxidans* catalyze electron transfer from animal ferrocytochrome *c* to NAD by means of ATP energy (Aleem *et al.,* 1963). The application of the fluorometric method of detecting reduced pyridine nucleotides showed that cells treated with ultrasound or frozen-thawed cells of *T. ferrooxidans* could reduce pyridine nucleotides at the expense of ATP energy and the electrons of $FeSO_4$ or ascorbate (Tikhonova *et al.,* 1967). The electron acceptor in the reverse transfer could be endogenic or added NAD. In the presence of uncouplers such as DNP and arsenate, NAD could not function as an oxidant of bacterial cytochromes.

How are oxidation and phosphorylation regulated in bacterial cells? They do not have the sensitive regulatory mechanisms found in mitochondria, *viz.,* respiratory control or regulation of oxidation of extramitochondrial NADH, to say nothing of higher forms of regulation, such as hormonal regulation (thyroxine, etc.).

Bacteria are not totally devoid of respiration-regulating mechanisms, however. A special mechanism has been found in fragments of *M. phlei* membranes: Adenosine monophosphate, formed by the action of adenylate kinase on ADP, stimulates NADH dehydrogenase, thus causing stimulation of the respiratory chain and the accumulation of ATP and NAD, which suppress the stimulation due to adenosine monophosphate (Worcel *et al.,* 1965). A possible mode of regulation of the energetics of bacteria is the adaptive synthesis of respiratory enzymes which ensure survival in various, sometimes very diverse, conditions. Owing to their great flexibility and capacity for adaptation bacteria have very "labile" respiratory chains. The latter vary in their set of components, and the relative amounts of the components. Sometimes one species has two or more electron-transport pathways, which are easily interchangeable. Most bacteria in anaerobic conditions can switch the respiratory chains to terminal acceptors other than oxygen, such as nitrite, hydroxylamine, sulfur compounds, or organic substances.

We have already discussed the effect of cultivation conditions on the respiration of *P. aeruginosa.* This species synthesizes nitrite oxidoreductase or cytochrome oxidase, depending on whether the medium contains nitrite or oxygen (Yamanaka, 1964; Azoulay, 1964). Both enzymes are terminal for the whole cytochrome-containing respiratory chain.

The switching of the respiratory chain from oxygen to nitrate respiration, with which cytochrome b_1 is specifically associated, has been thoroughly investigated in *E. coli* (Egami *et al.,* 1961; Itagaki and Taniguchi, 1959; Heredia and Medina, 1960). The adaptive nature of nitrate respiration, which exists along with oxygen respiration. has been investigated in *A. fischeri, P. fluorescens, H. parainfluenzae,* some micrococci, and many other species (Sadana and McElroy, 1957; Allen and Van Niel, 1952; Fewson and Nicholas, 1961*c*; White, 1962, 1963; Arima and Oka, 1965*a*; Prakash *et al.,* 1966; Naik and Nicholas, 1966; Downey, 1966). It has been shown that oxygen is a strong competitive inhibitor of nitrate respiration. The respiratory-chain components are the same for oxygen and nitrate respiration, apart from the terminal enzymes.

The ability of bacteria to convert easily to facultative anaerobiosis is regarded as evidence of the evolutionary antiquity of their origin (Gaffron, 1964; Yamanaka and Okunuki, 1964; Yamanaka, 1964). Bacteria can adapt themselves just as easily to different substrates, or electron donors, by adaptive synthesis of the initial components of the respiratory chain. For instance, the hydrogen bacterium *Hydrogenomonas facilis* contains hydrogenase as the initial enzyme of the chain. When cultivated on lactate, succinate, or malate, *H. facilis* becomes adapted to them and loses hydrogenase (Ruban, 1961; Packer and Vishniac, 1955; Packer, 1958*a, b*).

Among thiobacilli, *T. novellus* can grow on organic substrates as well as by oxidation of inorganic sulfur compounds to sulfate (Peck, 1962). Iron-oxidizing chemoautotrophs, *Ferrobacillus ferrooxidans,* for example, can also become adapted to growth on glucose (Remsen and Lundgren, 1963). The chemoautotrophic bacterium *Desulfovibrio* accomplishes sulfate respiration with the aid of cytochrome c_3. Cell-free preparations contain an enzyme which transfers an electron from cytochrome c_3 to sulfate (Egami *et al.,* 1961). The initial donor is hydrogen. Yet these bacteria can exist by obtaining energy from oxidation of lactate or pyruvate in the same respiratory chain (Postgate, 1952). An investigation of chemoautotrophic bacteria has revealed that very many of them can change from oxidation of inorganic compounds to oxidation of organic substances, but there must be a period of adaptation (Zavarzin, 1964).

Another significant feature of bacteria is the variation in the quantitative composition of respiratory-chain components. The set and number of cytochromes and the activity of NADH oxidase in *E. coli* depend on the cultivation conditions (Gray *et al.,* 1966). Interesting investigations connected with the composition of the res-

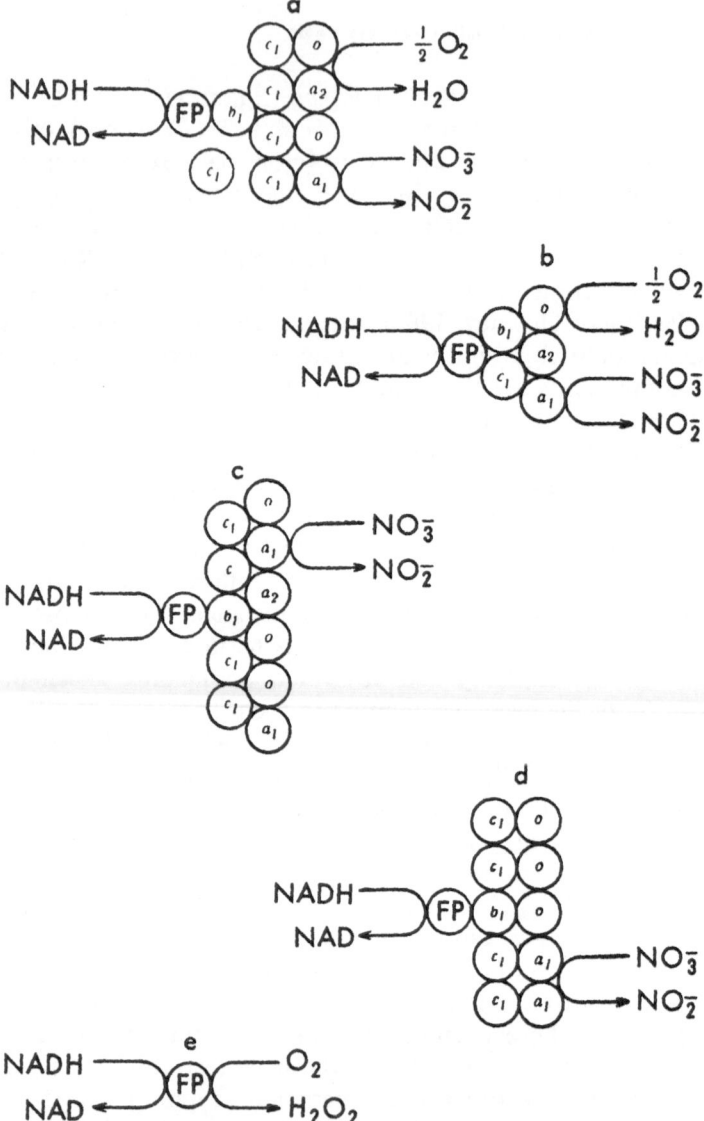

Fig. 18. Composition of respiratory chain of hemin-requiring *Hemophilus* sp. in different cultivation conditions. Growth conditions: (*a*) Levinthal's medium (aerobiosis); (*b*) proteose, peptone, 0.5 μg hemin per ml (aerobiosis); (*c*) proteose, peptone, 0.5 μg hemin per ml (CO); (*d*) proteose, peptone, 0.5 μg hemin per ml (anaerobiosis); (*e*) proteose, peptone, 0.002 μg hemin per ml, 0.5% glucose (aerobiosis). No respiratory chain enzymes were formed in media containing: proteose, peptone, 0.002 μg hemin per ml (aerobiosis); 0.002 μg hemin per ml, 0.5% glucose (anaerobiosis); or 0.5 μg hemin per ml, 0.005 M KCN (anaerobiosis) (from White, 1963).

piratory chain have been conducted with cells and cell-free preparations of *H. parainfluenzae,* the respiratory chain of which contains six cytochromes and five firmly bound flavin dehydrogenases. An interesting property of the respiratory chain of *H. parainfluenzae* is the high content of cytochrome c_1, weakly attached to the chain. Most of it washes off during disruption of the cell (White and Smith, 1962; Smith and White, 1962). The content of cytochromes, demethyl vitamin K_2, and dehydrogenases depends on the aeration and phase of growth of the culture (White, 1963, 1965*b*). This diversity of the enzyme composition of the respiratory chain and the variation in the amounts of individual components led White and Smith to conclude that in the bacterial respiratory chain the enzymes are not present in stoichiometric ratio, as Green postulated for mitochondria (Green and Hechter, 1965). The bacterial respiratory chain is probably more labile and consists of a three-dimensional mosaic of dehydrogenases and cytochromes, which easily change their composition and extent (White and Smith, 1964). As an example we can mention that in *H. parainfluenzae* the amount of D-lactate dehydrogenase can vary by a factor of 500, that of L-lactate dehydrogenase by a factor of 1400, the cytochrome a_1 content by a factor of 70, and the cytochrome c_1 content by a factor of 32–72 (Fig. 18).

Chapter IV

Oxidative Phosphorylation in Bacteria

GENERAL REMARKS

Oxidative phosphorylation, i.e., the formation of energy-rich adenosine triphosphate from the energy of substances oxidized in the organism, is generally recognized as one of the most important processes in the cells of animals and most microorganisms, since it ensures the accomplishment of almost all the other vital processes. The term "oxidative phosphorylation," as distinct from photosynthetic and glycolytic phosphorylation, means that the transfer of electrons (hydrogen) from the oxidizable substance to oxygen or another acceptor is effected by means of specific dehydrogenases and other enzymes forming the respiratory chain. This process is represented schematically in Fig. 19.

In cells of eukaryotes, which include all organisms except bacteria and blue-green algae, oxidative phosphorylation occurs in the mitochondria. This was the reason for the very great interest manifested in mitochondria by biologists. At present, mitochondria are probably the most thoroughly investigated of all the cell organelles. The mechanism of oxidative phosphorylation has been very intensively investigated in many laboratories of the world, but much more work is required before it is completely understood.

There are several hypothetical schemes representing the process of oxidative phosphorylation, but none of them give any information regarding the nature of the high-energy bond in compounds preceding the formation of ATP. The solution of this problem is now in view, however, mainly as a result of the work of Skulachev (1963, 1964*a*, *b*). Skulachev suggested that the pyrimidine ring and phosphate "tail" of the adenylic part of NAD and FAD molecules, and

161

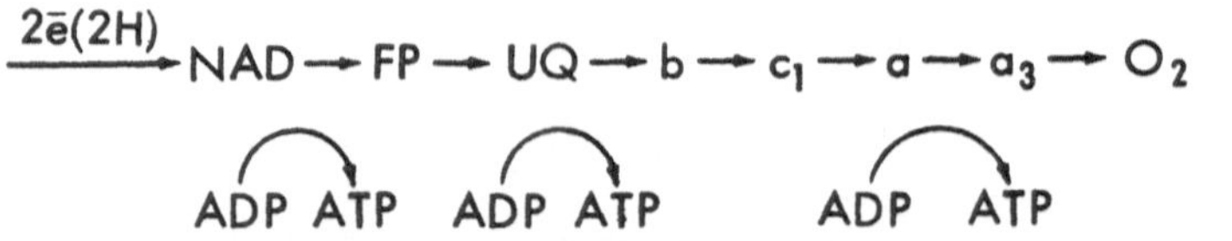

$$\xrightarrow{\text{2ē(2H)}} \text{NAD} \longrightarrow \text{FP} \longrightarrow \text{UQ} \longrightarrow \text{b} \longrightarrow c_1 \longrightarrow a \longrightarrow a_3 \longrightarrow O_2$$

ADP ATP   ADP ATP   ADP   ATP

Fig. 19. Schematic representation of oxidative phosphorylation in mitochondria.

bound internal adenosine diphosphate of the mitochondria, are most directly implicated in the formation of the high-energy bond. The literature contains excellent reviews on oxidative phosphorylation in mitochondria and their fragments, also covering a number of the related processes (Kotel'nikova, 1962, 1964; Skulachev, 1962; Zvyagil'-skaya, 1964; Ernster and Lee, 1964; Lehninger, 1964; Racker, 1965).

In this chapter we will discuss the process of oxidative phosphorylation in bacteria, since in bacteria it has special characteristics which may shed light on the mechanism of oxidative phosphorylation in general and on its evolution. This topic has already been discussed by Mitchell (1959a, b), Marr (1960), and Pinchot (1965). The scope of this review does not include analysis of photosynthetic phosphorylation in bacteria; this would be impossible without an analysis of the mechanism of photosynthesis as a whole. However, the mechanism of photosynthetic phosphorylation has been thoroughly examined by Sisakyan and Bekina (1964), and by Kondrat'eva (1964).

STRUCTURAL ORGANIZATION OF OXIDATIVE PHOS-PHORYLATION IN BACTERIA

Since the complex system of internal membranes in bacteria and many other structural features of the bacterial cell were discovered within the last five or six years, early investigators of oxidative phosphorylation in bacteria were obliged to work in the dark and to make do with coarse experiments. Extracts from *E. coli* (Pinchot and Racker, 1951; Hersey and Ajl, 1951), *A. vinelandii* (Hyndman *et al.,* 1953), *A. faecalis* (Pinchot, 1953), *M. phlei* (Brodie and Gray, 1955), and other bacteria were obtained by sonication of bacterial suspensions or by grinding with glass or aluminum oxide and removal of the cell fragments by centrifugation. Of course, nothing definite could be concluded regarding the localization of oxidative phosphorylation or the respiratory enzymes.

In the late fifties, a large number of investigations of bacterial phosphorylation was conducted, mainly in the laboratories of Slater,

Brodie, Marr, and Pinchot. The results of these investigations are discussed in the reviews of Mitchell (1959a, b) and Marr (1960). It was established that the respiratory chain is located in lipoprotein particles, which De Ley (1960) suggested should be called oxido-somes. These particles were precipitated by 30- to 60-min centrifuga-tion at 10,000–20,000 g when they were obtained by osmotic shock, or at 100,000–140,000 g after mechanical or ultrasonic disruption of the cells. Oxidative phosphorylation on *A. vinelandii* particles did not require any additional factors (Rose and Ochoa, 1956). Preparations from *P. vulgaris* and *M. phlei* showed enhanced oxidative phosphory-lation when the solution left after precipitation of the particles was added. Particles from *A. faecalis* phosphorylated only in the presence of a dissolved protein factor and a polynucleotide. Since information on the structure of the bacterial cell was meager it was assumed that the oxidosomes were derived from the cytoplasmic membrane.

The question of the localization of oxidative phosphorylation was complicated by the discovery of intracellular membranous struc-tures in bacteria. The entire argument relating to the participation of the bacterial mesosomes and cytoplasmic membrane in respiration (see Chapters I and III), the settlement of which entailed so much cytochemical and biochemical information, is to a large extent ap-plicable to the question of the localization of oxidative phosphoryla-tion (Ostrovskii and Gel'man, 1963; Ostrovskii, 1964b). Without fur-ther discussion of the cytochemical data, we mention some other approaches to this problem.

For instance, Feo *et al.* (1962) suggest that the sharp reduction in P/O ratio when normal *Proteus morganii* cells are converted to L forms or penicillin spheroplasts (from 1 to 0–0.2 and 0.5, respectively) is due to structural damage to the cytoplasmic membrane as a result of osmotic pressure. These authors considered their results evidence of the surface localization of respiratory enzymes. The discovery of a system of internal membranes in anaerobically grown microorgan-isms, such as *L. monocytogenes* and *Actinomyces bovis,* or in obligate anaerobes, even those which do not have cytochromes (Edwards and Stevens, 1963; Fitz-James, 1962) and the absence, on the other hand, of mesosomes in some aerobic organisms (Hines *et al.,* 1964) are also a strong argument against the implication of the mesosomes in elec-tron transport and oxidative phosphorylation.

Of great interest at present are the views developed by a group of researchers (Fernandez-Móran *et al.,* 1964) on the basis of the earlier observations of Fernandez-Móran (1963) and investigations in Green's laboratory (Green and Hatefi, 1961; Green, 1962c; Green et al., 1963). These investigators suggested, and to some extent demonstrated, that the respiratory chain and oxidative phosphorylation enzymes are localized in mushroom-shaped subunits, present in large numbers in each mitochondrion and situated on the surface of the cristae and internal mitochondrial membrane.

Recently Biryuzova et al. (1964) found similar subunits in a membrane preparation from M. lysodeikticus. Bladen et al. (1964) found them on membranes of Eubacterium sp., and Abram (1965) found them in eight other representatives of aerobic bacteria. In the first case, the subunits were situated on the internal membranes, which are evidently derived from the mesosomes, while in the second case they were found on the cytoplasmic membrane. Whatever the function of the subunits may be, their presence on the membranous structures of bacteria indicates that these structures perform mitochondrial functions. This is indirect evidence that oxidative phosphorylation takes place both in the cytoplasmic membrane and in the mesosomes.

Thus, despite the great sophistication of cytological, biochemical, and cytochemical methods, the question of the localization of the enzymes of respiration and oxidative phosphorylation in bacteria remains unresolved. We can assume, like Stanier (1964), that the enzymes of electron transport and oxidative phosphorylation are located in the cytoplasmic membrane itself and its derivatives—mesosomes of various types. However, the problem of the distribution of the enzymes will obviously remain unresolved until the various membranous structures can be preparatively separated.

OXIDATIVE PHOSPHORYLATION IN *ALCALIGENES FAECALIS*

Many bacteria, particularly A. faecalis, E. coli, M. phlei and A. vinelandii have been studied in the investigation of oxidative phosphorylation. Although mitochondria of different species of animals or different organs have in general the same or similar properties, this is not true of preparations from bacteria. In bacteria, oxidative phosphorylation not only differs from that in higher organisms, but is very variable in many characteristics (requirement of cofactors, sensitivity to inhibitors) even within a particular group of bacteria. Hence, it is more suitable to consider the mechanism of oxidative phosphorylation in relation to the organism under investigation.

The study of oxidative phosphorylation in animals began in 1939 with the investigations of Belitser and Tsybakova (1939), but bacteria did not attract attention in this respect until 1951, when Pinchot and Racker used an *E. coli* extract. After this, however, Racker gave up the study of bacterial phosphorylation altogether, while Pinchot switched over to *A. faecalis* and has been investigating it for more than 13 years. Pinchot's choice of organism (1953) was dictated by the need to suppress glycolysis in the preparation during the investigation of oxidative phosphorylation. *A. faecalis* does not ferment glucose and, hence, does not require the use of glycolytic inhibitors, which complicate the work.

The *A. faecalis* preparation, as Beer (1960) showed in his electron-microscopic investigation, consisted of lipoprotein-membrane fragments obtained by treatment of the bacterial cells with ultrasound (Pinchot, 1953). These fragments, together with the cell cytoplasm, effected oxidative phosphorylation in the oxidation of NADH, but after precipitation with ammonium sulfate, redissolving, and subsequent precipitation by centrifugation the particles no longer catalyzed oxidative phosphorylation without the addition of the supernatant solution and a hot-water extract of whole cells. The supernate factor was a protein and the factor in the hot extract was a polynucleotide of low molecular weight, probably a tetranucleotide containing uridylic, adenylic, and guanylic acids in the ratio 1:2:1 (Pinchot, 1955; Shibko and Pinchot, 1961b).

The polynucleotide was localized in the membranes but lost during their purification (Pinchot, 1957a). The natural polynucleotide could not be replaced by synthetic polyuridylic or polyadenylic acid, or by other polyanions (Shibko and Pinchot, 1961a). Judged by the nonincorporation of radioactive phosphorus, the polynucleotide is not an intermediate in oxidative phosphorylation (Pinchot, 1957c). Shibko and Pinchot (1961a) think that the polynucleotide and Mg^{++} (optimum concentration $4 \times 10^{-2}\ M$), bind the protein coupling factor to the particles. Mg^{++} is absolutely necessary, since without these ions even the complete set of factors cannot restore oxidative phosphorylation. Hovenkamp (1956b) suggests another explanation of the role of the polynucleotide. She thinks that the polynucleotide

can act as an accumulator of cations, the presence of which in considerable quantity, according to her data, stabilizes the membrane
and enzymes of oxidative phosphorylation.

However, the strength of the bond of the protein factor with the
membrane depended not only on the presence of Mg^{++} and the tetranucleotide. Pinchot (1960) observed that after incubation of the
complete phosphorylating system with substrate (NADH), but without P_{inorg} and ADP, the protein factor was separated from the membrane and was not precipitated with it on centrifugation. It was later
found that the protein factor, separated from the membrane in the
presence of DPNH, could bind inorganic phosphate and, in the
presence of ADP, transfer this phosphate to the terminal phosphorus of ATP (Pinchot, 1960; Scocca and Pinchot, 1963; Pinchot,
1963). The factor yielded equimolecular quantities of NAD. In other
words, in these experiments the protein factor behaved as an intermediate of oxidative phosphorylation and appeared to be a substance
of the protein $\sim$ NAD type. The authors envision the sequence of
reactions illustrated in Fig. 20 (Pinchot and Hormanski, 1962a, b;
Pinchot, 1965).

In proposing such a scheme, Pinchot drew attention to the following circumstances. First of all, the energy-rich compound of the
phosphorylation factor with the electron (hydrogen) carrier contains

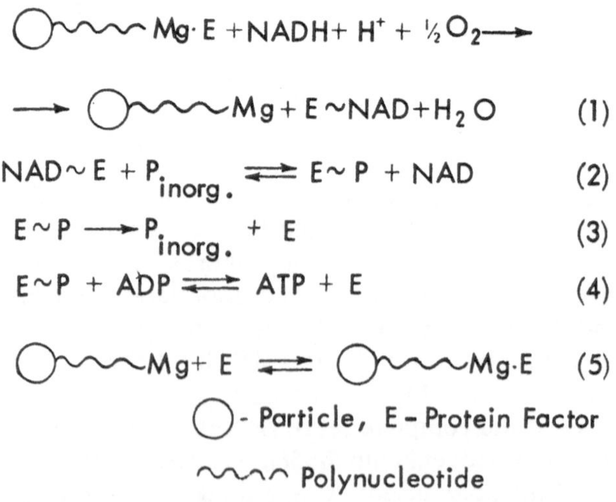

$$\bigcirc \!\!\!\sim\!\!\! Mg \cdot E + NADH + H^+ + \tfrac{1}{2} O_2 \longrightarrow$$

$$\longrightarrow \bigcirc \!\!\!\sim\!\!\! Mg + E \sim NAD + H_2 O \qquad (1)$$

$$NAD \sim E + P_{inorg.} \rightleftharpoons E \sim P + NAD \qquad (2)$$

$$E \sim P \longrightarrow P_{inorg.} + E \qquad (3)$$

$$E \sim P + ADP \rightleftharpoons ATP + E \qquad (4)$$

$$\bigcirc \!\!\!\sim\!\!\! Mg + E \rightleftharpoons \bigcirc \!\!\!\sim\!\!\! Mg \cdot E \qquad (5)$$

$$\bigcirc - \text{Particle}, \quad E - \text{Protein Factor}$$

$$\sim\!\!\sim\!\!\sim \text{Polynucleotide}$$

Fig. 20. Mechanism of oxidative phosphorylation in *Alcaligenes
faecalis,* according to Pinchot and Hormanski, 1962.

this carrier (NAD) in oxidized form. This is very important, since numerous investigations on mitochondria prior to Pinchot's work did not permit a definite answer to the question of the form of carrier in the energy-rich compound.

Secondly, this energy-rich compound (NAD $\sim$ E), leaves the site of its formation and passes into the solution. This possibly occurs in the living bacterial cell. This fact permits a different approach to the mechanism of "respiratory control," manifested in mitochondria as an inhibition or acceleration of respiration in the absence or presence of ADP. Chance and Williams suggest that in the case of ADP deficiency the electron carriers accumulate in an inactive state, as a factor $\sim$ carrier substance, so that respiration is inhibited or completely suppressed (Chance and Williams, 1955). Pinchot thinks that the state of respiratory control is the result of reversible removal of the carrier from the respiratory chain. The very plausible respiratory-control mechanism suggested by Pinchot cannot account for the almost complete absence of respiratory control in bacterial preparations, particularly in preparations of *A. faecalis* (Scocca and Pinchot, 1965).

Pinchot's work leads to a quite different concept of the mechanism of action of 2,4-dinitrophenol (DNP), a well-known and very effective uncoupler of oxidative phosphorylation in mitochondria. Pinchot showed that the presence of DNP prevents the return of the factor (E) which couples respiration to oxidative phosphorylation to membrane, where the oxidation of various substances occurs. Before Pinchot's work it was thought that DNP could catalyze the breakdown of an intermediate energy-rich compound without a phosphoryl group.

What are the characteristics of this process, the mechanism of which is now beginning to yield to investigation? Using NADH as a respiration substrate, Pinchot in his first investigation (1953) obtained the ratio $P/O = 0.4$; believing this low value to be accidental, he conducted a new series of investigations, where he found that P/O lay in the range 0.1–0.3 (Pinchot, 1957a). We recall that in oxidation of NADH in mitochondria a minimum of three ATP molecules per molecule of substrate is synthesized ($P/O = 3$) and, according to recent data, the ratio P/O can even reach a value of 5–6 (Smith and Hansen, 1962; Gurban and Cristea, 1965).

Even with the use of the most up-to-date optical technique of measuring oxidative phosphorylation, 1 was the maximum P/O ratio that could be found in a preparation of *A. faecalis*. In this method respiration was measured as the change in amount of NADH from the change in optical density at $\lambda = 340$ mμ, and phosphorylation was measured from the amount of NADPH (measured in the same way from ΔE_{340} mμ) formed from NADP in the presence of ATP and added glucose, hexokinase, and glucose-6-phosphate dehydrogenase (Pinchot, 1957d, b).

Oxidative phosphorylation in *A. faecalis* was found to be fairly insensitive to the uncoupling action of 2,4-dinitrophenyl. A DNP concentration of 7×10^{-5} M had no effect at all; a concentration of 2×10^{-4} M caused 60% uncoupling.

Although the illustrated scheme of reactions and the other data are a great step forward in the interpretation of the mechanism of oxidative phosphorylation, this is only the first step. The actual relative arrangement of all the factors and the precise role of each of them are still completely obscure. Why does the NAD$\sim$ E compound leave the particles which carry the respiratory chain enzymes? Why is ten times more DNP required to uncouple oxidative phosphorylation in this case than in animal mitochondria? Why does the P/O ratio not exceed unity in conditions where animal mitochondria give P/O = 3? Finally, why does an *A. faecalis* preparation not reveal a P_{32}–ATP exchange reaction? These and other questions necessitate a search for new ways of explaining the experimental data and raise doubt that the matter is understood (Pandit-Hovenkamp, 1965). We will return to these questions at the end of the review.

OXIDATIVE PHOSPHORYLATION IN *AZOTOBACTER VINELANDII*

Many investigators have tried to decipher bacterial oxidative phosphorylation by using *A. vinelandii,* an organism with a very active system of respiratory enzymes. This was first done by Hyndman *et al.,* (1953), who showed that both the particulate fraction and the supernatant fluid were required for oxidative phosphorylation (with H_2, glycerol, etc. used as substrates).

In their best experiments the P/O ratio did not exceed 0.24. In subsequent investigations by Rose and Ochoa (1956) and Cota-Robles *et al.* (1958), it was reported that oxidative phosphorylation on *A. vinelandii* particles does not require the soluble fraction, and Tissières confirmed that this process occurs in both fractions and is stimulated when they are combined (Tissières, 1954, 1956; Tissières and Slater, 1955; Tissières *et al.,* 1957).

This contradiction was explained in later investigations by Hovenkamp (1959*a, b*) and Pandit-Hovenkamp (1964), who found that *A. vinelandii* particles obtained by treatment with ultrasound or grinding with Pyrex glass powder carried all that was required for oxidative phosphorylation. When such particles were treated with solutions of low ionic strength, the unknown phophorylation factor was destroyed but could be replaced by a factor from the solution (obtained in the first precipitation of the particles) in the presence of a sufficient number of ions. Provisional information indicated that this factor was a protein.

Oxidative phosphorylation in *A. vinelandii* preparations, as in the case of *A. faecalis,* is distinguished by its very low efficiency, as can be seen from the P/O values given in Table 20.

The fact that many investigators have worked with *A. vinelandii* independently of one another rules out accidental results, and the difference in the results is obviously due to differences in the cultivation and treatment of the bacteria. The immense importance of cultivation conditions, phase of growth, and the rate and method of treatment of the bacteria in the investigation of oxidative phosphorylation has been shown in Zaitseva's works (1963, 1965).

Feo *et al.,* (1962), working with *P. morganii,* also noted that oxidative phosphorylation was considerably altered when normal bacterial cells were converted to penicillin spheroplasts or L forms. The P/O ratio fell from 1 to 0–0.2 and 0.5, respectively. The authors attributed such changes to morphological damage to the cytoplasmic membrane due to osmotic pressure.

Zaitseva investigated the most difficult, and probably the most interesting problem of oxidative phosphorylation, the process in the intact cell. Her predecessors (Rakestraw *et al.,* 1957) could not even detect oxidative phosphorylation (P/O = 0.001–0.003). Using the rapid incorporation of radioactive phosphorus into the cell and subsequent analysis of the first products to incorporate P_{32}, Zaitseva and her colleagues showed that the efficiency of oxidative phosphorylation reached a maximum at the time when the cells in a synchronous culture of *A. vinelandii* were preparing to divide (P/O up to 1.5). When the culture was cooled the P/O ratio dropped to 0.5 or lower (Zaitseva *et al.,* 1961; Zaitseva *et al.,* 1963). These authors showed that ATP, and no other compound, is synthesized in the

Table 20. Efficiency of Oxidative Phosphorylation in *Azotobacter vinelandii* Preparations

| Substrate | P/O | Effect of DNP on oxidative phosphorylation | | Author |
		DNP concentration, M	Uncoupling, %	
H$_2$, Glycerol	0.24	—	—	Hyndman *et al.*, 1953
Succinate	0.35	10^{-4}	35	Tissières, 1955
H$_2$	0.3	10^{-3}	30	Rose and Ochoa, 1956
Succinate	0.45	10^{-3}	80	
NADH	0.08	—	—	
Succinate	0.5	—	—	Tissières *et al.*, 1957
NADH	0.5	—	—	
Succinate	0.5	$8 \cdot 10^{-5}$	80	Hartman *et al.*, 1957
Malate	0.5	$8 \cdot 10^{-5}$	40	
H$_2$	0.2	—	—	Cota-Robles *et al.*, 1958
Succinate	0.45	—	—	Temperli *et al.*, 1958
Ketoglutarate	0.54	—	—	
NADH	0.5	10^{-3}	10	Hovenkamp, 1959*b*
NADH	0.2	—	—	Schils *et al.*, 1960
K$_3$H	0.2	—	—	
Succinate	0.7	$5 \cdot 10^{-4}$	50	Temperli *et al.*, 1960
Succinate	0.55	$5 \cdot 10^{-4}$	75	Kotel'nikova and
Glutamate	0.43	$5 \cdot 10^{-4}$	90	Ivanova, 1964

process of oxidative phosphorylation in *A. vinelandii*. This fact and the sensitivity to menadione (Schiels *et al.*, 1960) and oligomycin (Kotel'nikova and Ivanova, 1964) demonstrate the similarity with oxidative phosphorylation and animal mitochondria. At the same time, bacterial preparations do not exhibit any respiratory control, i.e., a relationship between the respiration rate and the presence of P$_{inorg}$ and ADP (Rose and Ochoa, 1956), or exhibit only very slight control (Kotel'nikova and Ivanova, 1964). Bacterial respiration is insensitive to antimycin (Schiels *et al.*, 1960) and phosphorylation is fairly insensitive to 2,4-dinitrophenol (see Table 21). This again brings out the special organization of oxidative phosphorylation in bacteria.

A comprehensive review of the literature on oxidative phosphorylation and other problems of metabolism in *A. vinelandii* can be found in Zaitseva's works (1963, 1965).

Table 21. Efficiency of Oxidative Phosphorylation (P/O) in Preparations from Various Bacteria (*E*–cell-free extract; *P*–subcellular particles; heavy (*h*) and light (*l*); *S*–cytoplasm solution)

Organism	Preparation	Succinic acid	Malic acid	NADH	Author
Escherichia coli	*E*	0.24	0.4	0	Kashket and Brodie,
	P	0.15	0.25	0.1	1963*a*
	$P_l + S$	0.5	1.1	0.3	
	$P_h + S$	0	0	—	
Azotobacter vinelandii	*E*	0.55	0.59	0.43	Kotel'nikova and
	P	0.87	0	0.41	Ivanova, 1964;
	P + S	0.87	0.41	0.32	Brodie, 1963
Mycobacterium phlei	*E*	1.78	1.8	1.5	Brodie, 1959
	P	0.56	0	0	
	P + S	0.98	1.6	1.1	
Aerobacter aerogenes	*E*	—	—	0.33	Nossal *et al.*, 1956
	P			0.16	
Corynebacterium creatinovorum	*E*	1.4	—	—	Brodie and Gray, 1956
	P	0.56			
Proteus vulgaris	*E*			0.62	Nossal *et al.*, 1956
	P	—	—	0	
	P + S			0.52	
Alcaligenes faecalis	*E*			0.36	Pinchot, 1957*a, b, c, d*
	P	—	—	0.03	
	P + S			0.33–0.78	
Micrococcus lysodeikticus	*E*		0.3	0.43	Ishikawa and Lehninger, 1962;
	P		0.1–0.65	0.13–0.73	Ostrovski and Gel'man, 1965
Proteus morganii	*E*	1	—	—	Feo *et al.*, 1962
L form	*E*	0.5			
Rhodospirillum rubrum	*P + S*	0.3	0.35	0.22	Geller, 1962; Taniguchi and Kamen, 1965
	P + S	0.1–0.4	—	0.1–0.4	
Gluconobacter liquefaciens	*E*	0.14	—	0.09	Stouthamer, 1962
	P	—	—	0.13	
Bacillus subtilis	*P + S*	—	1.85	—	Downey, 1964
Nitrobacter winogradskii	*P*	—	—	2.2–2.8	Kiesow, 1964
Streptococcus faecalis	*E*	—	—	0.2	Gallin and Van Demark, 1964
Brevibacterium Ketoglutaricum	*E*	—	1.6–2.3	—	Zeng-Yi Shen and Chow Kwang-yu, 1965

OXIDATIVE PHOSPHORYLATION IN *ESCHERICHIA COLI*

There have been a few investigations of oxidative phosphorylation in *E. coli*. Incidentally, this bacterium opened up the era of investigation of bacterial oxidative phosphorylation, when Hersey and Ajl and Pinchot and Racker in 1951 found this process in *E. coli* extracts with a P/O ratio equalling 0.55 and 0.9, respectively, for the oxidation of succinate and ethanol.

It was later found that ultrasonic particles from *E. coli* were incapable of oxidative phosphorylation (Asnis *et al.*, 1956). The attempts of such well-known investigators as Kashket and Brodie (1963*a*) to fractionate cell-free homogenates of *E. coli* produced a large amount of new information. These authors separated the ultrasonic homogenate of the bacteria into a heavy-particle fraction (precipitated in 45 min at 39,000 *g*) and a supernate. The heavy particles carried succinic oxidase and were incapable of oxidative phosphorylation even when the supernate was added. The supernate oxidized NADH without coupled phosphorylation. Only the light-particle fraction together with supernate proteins extracted with ammonium sulfate effected oxidative phosphorylation on various substrates. The P/O ratios were 0.5, 1.0, 1.0, and 0.3, respectively, when succinate, α-ketoglutarate, malate, and NADH were used as the oxidation substrate. The addition of the supernate to the particles increased the respiration rate and absorption of P_{inorg}. Oxidative phosphorylation on the light particles was very unstable and disappeared when they were washed, but was not altered by the addition of 2,4-DNP, amobarbital, or antimycin to the particles. Pinchot and Racker (1951) also observed insensitivity to 2,4-DNP, but Hersey and Ajl (1951) reported that 6×10^{-5} *M* DNP (like gramicidin) caused 60% uncoupling of oxidative phosphorylation, while Nisman *et al.*, (1963) claimed that 5×10^{-5} *M* DNP caused 30% uncoupling. Ota *et al.*, (1964) reported recently that when NO_3^- was used as an electron acceptor, 10^{-4} *M* 2,4-DNP caused 50% inhibition of oxidation, reducing P/O to 30%, while amobarbital (10^{-3} *M*) and antimycin (10^{-3} *M*), though not affecting nitrate reduction, reduced P/O by 90 and 80%, respectively.

The very contradictory results are unexpected and must be attributed to differences in the experimental conditions. That such differences are significant is indicated by Rose and Ochoa's data (1956) showing that at pH 7 10^{-3} *M* DNP caused 30% uncoupling of oxidative phosphorylation in an *A. vinelandii* preparation (oxidation substrate H_2), while at pH 6 the uncoupling was 98%. The phosphorylation mentioned above, that is associated with electron transfer from the substrate to NO_3^-, also occurs on *E. coli* particles only in the presence of soluble cofactors (Yamanaka *et al.*, 1964; Ota *et al.*, 1964). The efficiency of this nitrate respiration is given by the ratio $P/NO_3^- = 0.1$–0.65 (Hempfling *et al.*, 1964; Ota *et al.*, 1964). Using gel filtration and ion-exchange chromatography on DEAE cellulose columns, Ota (1965) managed to separate the phosphorylation factor into several active fractions (F_1, F_2, and F_4), in which the coupling activity was not correlated with the presence of ATPase and the ATP–P_{32} exchange reaction. The incorporation of inorganic phosphate into the molecule of the most active factor (F_4) was 65% inhibited by 2,4-DNP in a concentration of 10^{-3} *M*, but phosphoryl transfer from F_4 to ADP was unaffected. The coupling power of a mixture of all three factors was very much greater than that of factor F_4 alone.

Nisman *et al.*, (1963) found that NADH oxidation, which could only be effected by *E. coli* particles obtained from treating penicillin spheroplasts treated with digitonin, was very strongly inhibited by the addition of ATP and various nucleoside diphosphates.

Thus, an investigation of oxidative phosphorylation in *E. coli* again accentuates the difference between bacterial phosphorylation and the classical mitochondrial phosphorylation of higher organisms. However, Zeng-I-shen (1965) recently showed that cell-free extracts of *E. coli* can effect oxidative phosphorylation with an efficiency (P/O = 2) comparable with that of mitochondria if the bacterial preparations are obtained in the presence of considerable amounts of albumin and Mg^{++}.

OXIDATIVE PHOSPHORYLATION IN *MYCOBACTERIUM PHLEI*

Brodie and his colleagues achieved considerable success in work with *M. phlei* preparations. Beginning with an investigation of coarse cell-free extracts, obtained by ultrasonic treatment of the cells (Brodie and Gray, 1955), and finding efficient oxidative phosphorylation in these extracts (P/O = 1.2–2.0 on various substrates, except lactate and NADH) (Brodie and Gray, 1955, 1956*a, b*), they soon established that both phosphorylation and oxidation require the presence of particles and supernate (Brodie and Gray, 1957; Asano and Brodie, 1963; Asano *et al.,* 1964; Brodie, 1959). The particles contained all the respiratory enzymes and could independently oxidize lactic acid. Nevertheless, the oxidation of other substrates required some factor, obviously protein, from the supernate. The reason why this factor was not separated from the phosphorylation factor during fractionation of the supernate with chemical agents is still a mystery.

Simultaneously with the discovery of the supernate factors in Brodie's laboratory, it was noted that a short exposure of *M. phlei* particles and solution to ultraviolet light ($\lambda = 360$ mμ) suppressed phosphorylating ability but did not affect the activity of the respiratory enzymes. Substrate oxidation was suppressed by more intense illumination (Brodie *et al.,* 1957). Oxidizing ability was restored by the addition of flavin nucleotides (FMN), but oxidation and phosphorylation could both be restored only by the addition of vitamin K_1 (Brodie *et al.,* 1958; Weber *et al.,* 1958; Brodie, 1961*b*).

The authors suggested that vitamin K_1 is both a component of the electron-transport chain and a participant in the production of energy-rich phosphate. They actually managed to extract a naphthoquinone of the vitamin K_1 type and a phosphorylated derivative capable of transferring energy-rich phosphate to ADP with the aid of the *M. phlei* enzyme system (Kashket and Brodie, 1960; Russel and Brodie, 1960; Weber *et al.,* 1963; Brodie, 1961*b*)

Similar observations on the regeneration of the respiratory

chain by means of vitamin K derivatives have also been made for *A. xylinum* (Benziman and Perez, 1965), *H. parainfluenzae* (White, 1965a), and *B. subtilis* (Downey, 1964).

Taking into account the special features of the structure of vitamin K_1 and its known properties, Brodie and Ballantine (1960) proposed a scheme for the processes occurring in oxidative phosphorylation in *M. phlei* with the participation of vitamin K.

This scheme agrees with the ideas of other authors (Dallam, 1961) regarding the possible formation of a high-energy bond in the vitamin K_1 molecule, although there are some differences in details. Scott (1965), however, from an investigation of synthetic chromanyl phosphates and quinones is very dubious of their ability to take part in oxidative phosphorylation.

Brodie and Ballantine (1960) suggest that the tendency to form a high-energy bond is due to the ability of vitamin K_1 to form a ring and produce chroman. This hypothesis was confirmed in their work in which they used various vitamin K_1 analogs substituted in positions 2 and 3, and in experiments with the 6-chromanyl phosphate of vitamin $K_{1(20)}$ (Asano *et al.*, 1962). Phosphorylating preparations from *M. phlei* can incorporate tritium into vitamin K_1. This reaction occurs in the presence of P_{inorg} and leads to the formation of 6-chromanyl phosphate. This again, though indirectly, indicates the implication of vitamin K_1 in oxidative phosphorylation (Gutnick and Brodie, 1965). Strictly speaking, the naphthoquinone from *M. phlei* is a derivative of vitamin K_2, not K_1 (it contains 45 carbon atoms in the side chain as against the 20 in vitamin K_1), but this does not invalidate the relationships discovered by means of vitamin K_1 and its derivatives (Weber and Rosso, 1963).

A soluble enzyme system which effects phosphorylation coupled with oxidation of vitamin K_1 was recently obtained in Brodie's laboratory (Watanabe *et al.*, 1965; Brodie and Adelson, 1965). This system is still very complex. It contains malate:vitamin K reductase, a protein coupling factor, phospholipid, FAD, vitamin K_1, MTT, ADP, and P_{inorg}, but it represents the possibility of successful reconstruction of oxidative phosphorylation from relatively simple components.

The work of Brodie and his colleagues on the role of naphthoquinones fit in with the ideas of Green and other researchers on the role of quinones in the energy metabolism of mitochondrial organisms (Green and Hatefi, 1961; Green, 1962; Hatefi, 1963; Hatefi *et al.*, 1961). These studies gave rise to a whole series of investigations on the occurrence of benzoquinones and naphthoquinones in the bacterial kingdom (see Chapter III).

M. phlei is exceptionally (for bacteria) sensitive to uncouplers of oxidative phosphorylation. For instance, $8 \times 10^{-5} M$ DNP inhibits phosphorylation by 79%, gramicidin ($7 \times 10^{-5} M$) causes 43% inhibition, thyroxine ($4 \times 10^{-5} M$) causes 61% inhibition, and Dicumarol ($7 \times 10^{-5} M$) causes 67% inhibition. Having investigated very thoroughly the electron transport pathways in the respiratory chain of *M. phlei*, Asano and Brodie (1964, 1965a) concluded that the comparatively low efficiency of oxidative phosphorylation in bacteria was due to the simultaneous functioning of several, often intersecting, pathways of electron transfer, only some of which, for obscure reasons, are coupled with ATP synthesis.

OXIDATIVE PHOSPHORYLATION IN *MICROCOCCUS LYSODEIKTICUS*

Ishikawa and Lehninger (1962) made very interesting observations on *M. lysodeikticus*. By combining lysis with lysozyme with ultrasonic treatment, these authors obtained particles capable of oxidative phosphorylation (P/O on NADH was 0.5 on the average) extremely resistant to the action of classical respiration inhibitors and uncouplers. For instance, 2,4-DNP ($10^{-3} M$), sodium azide, Dicumarol, oligomycin, antimycin, and KCN had hardly any effect on respiration or phosphorylation. Respiration was only 17% inhibited by $1.2 \times 10^{-3} M$ KCN and 24% inhibited by 10 μg/ml of oligomycin (a dose ten times the maximum dose for suppression of phosphorylation in mitochondria). Arsenate, menadione, and pentachlorophenol, however, all had a strong uncoupling effect.

The second special feature of particles from *M. lysodeikticus* is the ability to accomplish the "partial reactions" of oxidative phosphorylation: exchange of phosphorus between P_{inorg} and ATP (P—ATP exchange), exchange between C_{14} and ADP and ATP (ADP–ATP exchange), and the ATPase reaction. When the particles are put into a hypotonic solution, the enzymes of these reactions are dissolved and the particles become incapable of oxidative phosphorylation if no solution is added to them. When the coupling factor is removed only ADP–ATP exchange remains in the solution, but the P–ATP exchange and ATPase reactions reappear when the factor is added to the particles. The factor coupling oxidation with phosphorylation between ubiquinone and cytochrome c in animal mitochondria behaves in a similar way (Beyer, 1964).

Oxidative phosphorylation in *M. lysodeikticus* particles is greatly stimulated by a heat-stable protein factor, irrespective of the presence of the labile factor (Yamashita and Ishikawa, 1965). The heat-stable factor is not capable of the partial exchange reactions.

We think that the existence of exchange reactions in a solution outside the respiratory chain can best be explained by the following scheme of oxidative phosphorylation proposed by Slater and Hulsman (1959), where A and B are electron carriers, Y and X are unknown participants in the reactions; ADP–ATP exchange is a reflection of reaction (4), P–ATP exchange is a reflection of reactions (4) and (3), and ATPase activity is the reversal of reaction (3) with the breakdown of substances $X \sim P$ or $X \sim Y$.

$$AH_2 + B + Y \rightleftharpoons A \sim Y + BH_2 \qquad (1)$$

$$A \sim Y + X \rightleftharpoons A + Y \sim X \qquad (2)$$

$$Y \sim X + P_{inorg} \rightleftharpoons X \sim P + Y \qquad (3)$$

$$X \sim P + ADP \rightleftharpoons X + ATP \qquad (4)$$

If this is actually the case, the loss of P–ATP exchange and ATPase activity with the preservation of the ADP–ATP exchange reaction can be attributed to precipitation of substance Y on purification of the factor. In other words, ADP–ATP exchange indicates that the compound $X \sim P$ normally functions in the factor, and further transformations to $Y \sim X$ (necessary for the P–ATP exchange reaction) require only the presence of substance Y. How then are the lost reactions restored when the factor is added to the membrane particles? The substance Y does not remain in the particles either! It seems a logical assumption to us that X and Y form a single complex and are not separated by coarse fractionation of the solution, but only change their relative positions (Fig. 21) so that $X \sim P$ becomes incapable of interacting with Y. When the factor is returned to the membrane, spatial limitations compel X and Y to come together again, making reaction (1) and (2) possible.

Thus, from the results of Ishikawa and Lehninger (1962) and Yamashita and Ishikawa (1965) we can represent the set of effects occurring in a *M. lysodeikticus* preparation in the following way (see Fig. 21).

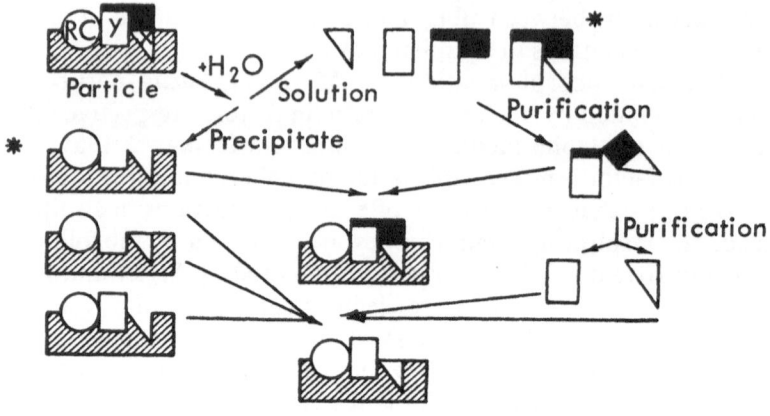

✱ Predominant Component of Mixture

Fig. 21. Schematic representation of reconstruction of oxidative phosphorylation in *Micrococcus lysodeikticus* membranes from results of Ishikawa and Lehninger, 1962; and Yamashita and Ishikawa, 1965.

When fragments of bacterial membranes are treated with hypotonic solutions the bulk of the membranes lose all their phosphorylation factors, but some of the fragments lose only individual factors (Y or X). Presumably Y is heat-stable, and X is heat-labile. Complexes of phosphorylation factors of different complexity remain in the solution. Fractionation of this mixture leads to the extraction of a complex of factors X and Y capable of effecting ADP-ATP exchange and of almost completely reconstructing oxidative phosphorylation in the membranes, but incapable of the other partial reactions owing to the inability of the factors to interact with one another. With further purification of the complex YX, its components are separated; each exhibits independently the ability to stimulate oxidative phosphorylation in membrane fragments. The effect of the mixture of factors naturally is much stronger than the sum of the effects of the individual factors.

It is of interest that in an *A. faecalis* preparation (Pinchot and Hormanski, 1962*a*) ATP–P exchange is absent, despite the demonstrated possibility of reversal of ATP synthesis. This reversal does not occur in the presence of P_{inorg}. Obviously, reaction (3) is shifted far to the right in this case.

Finally, one other feature of a preparation of *M. lysodeikticus* particles is that it requires the presence of a large number of magnesium ions (optimal concentration 0.01 M), although these can be replaced by spermine or spermidine. Respiratory control is absent in *M. lysodeikticus* particles. An interesting problem arises in connection with the observation of Ishikawa and Lehninger (1962) that Mg^{++} can be replaced by spermine or spermidine as a stabilizer of oxidative phosphorylation. Evidently Mg^{++} and these polyamines somehow assist the interaction of the factor and the membrane carrying the respiratory chain.

Spermine and spermidine are widely occurring components of bacterial cells; they can prevent osmotic shock of *M. lysodeikticus* protoplasts and *E. coli* spheroplasts in very low concentrations (on the order of 10^{-3} M), which magnesium ions cannot do, and they prevent the solution of *P. striata* membranes by EDTA (Tabor, 1962; Adams and Newberry, 1961). The stabilizers perhaps form a bridge between two structures or neutralize the negative charge on the membrane or on the factor, thus altering or preserving the configuration of the structures. In either case the question remains: What do Mg^{++} and, for instance, spermidine $[H_2N—(CH_2)_3—NH—(CH_2)_4—NH_2]$ have in common that enables them to have the same effect on bacterial oxidative phosphorylation, when many other compounds also carrying positive charges are ineffective?

On the basis of the above information and our own investigations, we think that phosphorylation factors are connected with the membrane by means of hydrogen bonds, and Mg^{++} or spermidine merely neutralizes the negatively charged groups on the membrane and factor, thus assisting their approach and interaction. Urea solutions even in the presence of large numbers of Mg^{++} in Tris-HCl buffer cause complete (but reversible) separation of the phosphorylation factors from the membranes. Treatment of phosphorylating preparations of *M. lysodeikticus* with bacterial proteinase, according to our results, leads to uncoupling of oxidative phosphorylation. This probably indicates the lack of a connection between protein factors and lipids which could protect them from digestion (Ostrovskii and Gel'man, 1965). Confirmation of this is provided by the data of Yamashita and Ishikawa (1965), who found inactivation of the heat-stable factor after treatment with proteolytic enzymes, and by our own observations, which showed that respiratory-chain enzymes are not inactivated by treatment with various proteinases (Ostrovskii *et al.*, 1964).

From the data of Pinchot, Brodie, and Ishikawa we can represent oxidative phosphorylation in bacteria by the following scheme, where C^{Co}, C' are carriers of electrons and coenzymes, and YX is the complex of the two phosphorylation factors.

$$C^{Co}_{red} + C'_{ox} + YX \rightleftarrows C^{Co}_{ox} \sim YX + C'_{red} \qquad (1)$$

$$C^{Co}_{ox} \sim YX \rightleftarrows C^{Co}_{ox} + \widetilde{YX} \qquad (2)$$

$$\widetilde{YX} + P_{inorg} \rightleftarrows YX \sim P \qquad (3)$$

$$YX \sim P + ADP \rightleftarrows ATP + YX \qquad (4)$$

Thus, the chain of transfer of the high-energy bond can be written briefly as

$$C^{Co} \sim YX \rightarrow \widetilde{YX} \rightarrow YX \sim F \rightarrow F \sim ADP$$

and the phosphorylation factors can then be represented as the complex YX for *M. lysodeikticus,* as $Co \sim YX$ (where Co is NAD) for *A. faecalis,* and as CYX for *M. phlei.*

SPECIAL FEATURES OF OXIDATIVE PHOSPHORYLATION IN OTHER BACTERIA

A special feature of bacterial oxidative phosphorylation is the ability of some bacteria to use mineral substances as respiration substrates (Zavarzin, 1964; Nicholas, 1964) or as terminal electron acceptors. For instance, *E. coli, P. denitrificans, P. aeruginosa* (Fewson and Nicholas, 1961*b*) and *M. denitrificans* (Hori, 1961; Fewson and Nicholas, 1961*a*; Naik and Nicholas, 1966) can transfer electrons to nitrate and in so doing effect the synthesis of energy-rich phosphates. Ohnishi and Mori (1960, 1962), Ohnishi (1963), and Yamanaka *et al.,* (1962, 1964) showed that this process occurs in ultrasonic fragments of *P. denitrificans* and *P. aeruginosa* membranes in the presence of soluble factors during the oxidation of NADH, lactate, and succinate. The ratio P/NO_3^- is 0.2–0.3, although E_0' of the NO_2^-/NO_3^- system is $+0.421$ v, i.e., the expected value for P/NO_3^- is 2. Ohnishi again asserts that the value of 0.2–0.3 is obtained for the case of succinate oxidation, and that the oxidation of NADH or lactate gives figures lower by a factor of 2–3. An *E. coli* preparation in similar conditions phosphorylates more vigorously with $P/NO_3^- = 0.65$ (Yamanaka *et al.*, 1964), and a preparation from *M. denitrificans*, with $P/NO_3^- = 0.5$ (Whatley, 1962). The sensitivity of the system to DNP is very low, and uncoupling requires concentrations of about 1×10^{-3} *M*.

However, despite the abundance of data indicating the low efficiency of oxidative phosphorylation in nitrate-reducing bacteria, which was beginning to appear a regular feature, some authors have managed to obtain bacterial preparations with very efficient phosphorylation. For instance, Kiesow (1964) showed that *N. winogradskii* has a NADH :NO_3^- oxidoreductase system which produces 1.2–2 ATP molecules per NADH molecule oxidized. On the basis of indirect data, Hadjipetrou and Stouthamer (1965) believe that intact *A. aerogenes* cells synthesize three ATP molecules per nitrate molecule reduced to nitrite.

The bacterium *N. europea* oxidizes hydroxylamine to nitrite with $P/2\bar{e} = 0.2$ (Ramaiah and Nicholas, 1964; Aleem, 1966*a*), while *N. agilis* uses nitrite as an oxidation substrate (Aleem and Nason, 1960; Malavolta *et al.*, 1960) and transfers electrons via the chain:

$$NO_2^- \rightarrow Cyt\ c \rightarrow Cyt\ a \rightarrow O_2$$

The synthesis of energy-rich phosphate ($P/O = 0.2$) occurring in this case can occur in the fraction of membrane fragments without the addition of soluble factors. As in many other bacterial systems, phosphorylation is insensitive to 5×10^5 *M* DNP, thyroxine, and Dicumarol. Oxidation is inhibited only by very high (for this poison) concentrations of antimycin A (20 μg/ml).

Kiesow (1964) suggests that nitrite oxidation is effected in two stages, one of which involves energy consumption and leads to the reduction of NAD, and the second involving oxidation of NADH by oxygen with the participation of the usual chain of electron (hydrogen) carriers.

Aleem (1965, 1966*a, b*) showed the reversal of electron transport in the respiratory chains of *N. agilis, N. europea*, and *T. novellus*. It is of interest that the reduction of NAD at the expense of thiosulfate electrons and ATP energy is very efficient: the ratio $ATP_{consumed}/NAD_{reduced}$ reaches 1. It might be expected that in the case of direct electron transfer the ratio $P/2\bar{e}$ is 1 only in this region, but this has not yet been found. On the contrary, Hempfling and Vishniac (1965) showed that oxidation of NADH, sulfide, sulfite, and thiosulfate by cell-free extracts of *Thiobacillus* X is not accompanied by phosphorylation at all; only the oxidation of mercaptoethanol gives $P/O = 0.44$.

Doman and Tikhonova (1965) have thoroughly discussed questions of the energetics of lithotrophs, i.e., organisms which use mineral oxidation substrates. We will mention here only the works of Fisher and Laudelout (1965*a*, *b*), who showed on cell-free extracts of *N. winogradskii* that the oxidation of NO_2^- by oxygen is accompanied by synthesis of ATP ($P/O = 0.13$–0.17). An analysis of the energy consumption of the cell and a calculation of Y_{ATP} (see below) indicates that such a low P/O ratio is characteristic of the intact cell too.

Of great fundamental interest is the discovery of oxidative phosphorylation in the ordinary anaerobe *S. faecalis* (Gallin and Van-Demark, 1964). Cell-free extracts of this bacterium in aerobic conditions effect the oxidation of NADH with $P/O = 0.2$, displaying in this an unusual sensitivity of the phosphorylation to 2,4-DNP (8×10^{-5} *M* DNP completely uncouples oxidation and phosphorylation). The authors note that when the bacterium is transferred to aerobic conditions it remains devoid of cytochromes, as before.

MECHANISM OF BACTERIAL OXIDATIVE PHOSPHORY-LATION IN GENERAL

What picture do we have of bacterial oxidative phosphorylation as a whole? Tables 20 and 21 show that this process is characterized by low efficiency and low sensitivity to uncouplers of the 2,4-DNP type. To this must be added lack of respiratory control, the comparatively easy fractionation, and the ability to use mineral substrates as electron acceptors (Table 21).

In mitochondria the only product of oxidative phosphorylation is ATP, but this question cannot be regarded as solved for bacteria. By using P_{32}, Zaitseva *et al.* (1961) showed that *A. vinelandii* synthesizes ATP before other energy-rich phosphates, but in *A. suboxydans* (Klüngsoÿr *et al.*, 1957) P_{32} is incorporated first into pyrophosphate ($P \sim P$), and in *B. cereus* spores (Srinivasan and Halvorson, 1961) ATP and IMP are labeled equally.

Baltscheffsky and von Stedingk (1966) made some very interesting observations in an investigation of photophosphorylation in bacteria. These authors found that, in the absence of adenine nucleotides, photophosphorylation led to the formation of pyrophosphate, which is possibly a precursor of ATP.

Senez (1962) reports that the absence of respiratory control can be found even in intact bacterial cells. For instance, when the synthetic processes are stopped for some reason and ATP consumption ceases, respiration continues as before. Senez thinks that the excess ATP is hydrolyzed by powerful ATPases in bacteria, but one can postulate the simple absence of ATP synthesis, i.e., oxidation of substrates by a free path.

In trying to explain the very obvious difference between bacterial oxidative phosphorylation and that in animal and plant mitochondria specialists have put forward several hypotheses:

1. Destruction of the bacterial cell leads to irreparable damage to the mitochondria-like structures.

2. The preparations contain powerful ATPase and ADPase, which distort the picture of synthesis of energy-rich phosphate.

3. All the above features are an integral part of bacterial oxidative phosphorylation.

Although bacterial preparations actually contain active ATPases in the membranes or cytoplasm (Voelz, 1964; Drapeau and Mac-Leod, 1963; Weibull et al., 1962), the use of fluoride, which inhibits their action, and hexokinase, which competes with them for the substrate, considerably reduces the risk of artifacts.

Destruction of the cell actually has a very pronounced effect on the state of the cytoplasm and membranes. Edebo (1961), who made a special study of this, reported that slight changes in pH, ionic strength, and other physicochemical conditions could lead to radical, often irreversible, changes in the structure of the bacterial cell. In view of this the data obtained for intact cells of A. vinelandii (Zaitseva et al., 1961, 1963), B. megaterium (Mal'tseva, 1963), A. suboxydans (Klüngsoÿr et al., 1957), and P. denitrificans (Ohnishi and Mori, 1960) are of particular interest. The P/O ratio of such "intact" oxidative phosphorylation was slightly greater than unity only in Zaitseva's work. Stouthamer (1962) obtained interesting data from a study of the degree of utilization of substrates by bacteria. He worked with a culture of G. liquefaciens growing in a medium with glucose. He found that from every ten glucose molecules the bacterium assimilated 4.5 molecules and 5.5 were oxidized to CO_2 and H_2O. The author thinks that, if 9 ATP molecules are spent on glucose assimilation and

5.5 ATP molecules are spent on the initial stages of oxidation of 5.5 glucose molecules, the total energy expenditure of the cell is 14.5 molecules obtained by oxidation of 5.5 glucose molecules, i.e., the oxidation of one glucose molecule gives 2.6 ATP molecules. Since the oxidation of one glucose molecule requires six oxygen molecules $(C_6H_{12}O_6 + 6O_2 \rightleftarrows 6CO_2 + 6H_2O)$, $P/O = \dfrac{2.6}{12} = 0.22$. Of course, the entire energy consumption of the cell is not taken into account (Dawes and Ribbons, 1964) and the P/O ratio is rather underestimated, but the data cited distinctly show the low efficiency of utilization of substrate oxidation energy by bacteria and that a low P/O is inherent to bacterial phosphorylation and not a preparation artifact.

The quantity Y_{ATP} (increment of dry mass of bacteria in grams per mole of ATP synthesized) (Senez, 1962; Oxenburgh and Snoswell, 1965) is being used more and more widely in microbial physiology. This quantity in itself does not give any indication of the efficiency of oxidative phosphorylation. For instance, for the yeast *Saccharomyces cerevisiae* and for the bacterium *S. faecalis,* $Y_{ATP} = 10.5$ (Senez, 1962), for *Torula utilis* (yeast), $Y_{ATP} = 10.9$ (Chen, 1964), and for *G. liquefaciens* (bacterium), $Y_{ATP} = 10$ (Stouthamer, 1962).

This value, however, calculated for a substrate such as glucose, which gives a known yield of ATP, can be used for an indirect determination of P/O on other substrates by a comparison of the increment of dry mass and the amount of substrate consumed. Investigations in this direction are just beginning, but the results already obtained are very interesting. For instance, Hadjipetrou *et al.* (1964) and Hadjipetrou and Stouthamer (1965), using this method, showed that intact cells of *A. aerogenes* utilize oxidation substrates very efficiently in aerobic conditions ($P/O = 3.1$) and when nitrate is used as an electron acceptor ($ATP/NO_3^- = 3$).

On analyzing the facts we have discussed, we are obliged to conclude that, on one hand, destruction of the cells is not the main reason for the special nature of bacterial phosphorylation and, on the other, that oxidative phosphorylation in intact cells is a fairly variable process, the efficiency of which depends on the species of bacterium and the cultivation conditions. We have mentioned already some recent works giving experimental and theoretical evidence of the presence of six points at which oxidation is coupled with phosphorylation in mitochondria (Smith and Hansen, 1964; Bieber, 1964; Lynn and Brown, 1965; Lenaz and Beyer, 1965a). If the values of redox potentials given for respiratory-chain components in the literature are correct this would seem improbable.

Values of P/O up to 7 have been reported even for bacterial preparations (Karpenko et al., 1964), but these data were obtained for anaerobic organisms (lactic acid bacteria), investigation of which in well-aerated conditions can damage their phosphorus metabolism. In later investigations (Lenaz and Beyer, 1965b; Koivusalo et al., 1966), it was shown that overestimated values of P/O were obtained in short experiments conducted in manometric vessels owing to the incomplete equilibrium between the liquid and gas phases, so that values of ΔO lower than the actual values, were recorded.

If we accept as fact the imperfect organization of bacterial phosphorylation, we must then explain it somehow in chemical and anatomical terms. The bacterial respiratory chain is regarded by most researchers as very similar to that of higher organisms, but bacteria abound in all kinds of deviations from the classical chain, especially as regards cytochromes, as shown in the previous chapter.

Some bacteria, e.g., M. lysodeikticus, which possess a cytochrome-containing respiratory chain, are insensitive to cyanide. Others, depending on conditions, switch easily from one oxidase to another (White, 1962; Castor and Chance, 1959). We think that this disposes bacteria, to a much greater extent than higher organisms, to utilize "free-oxidation" pathways (Skulachev, 1962), the energy of which is dissipated as heat, or is used in a different form from ATP, as, for instance, in the incorporation of C_{14} amino acids into mitochondrial proteins (Bronk, 1963; Kroon, 1964), in activation of fatty acids (Wojtczak et al., 1965), and in ion transport (Judah and Ahmed, 1964; Rossi and Lehninger, 1964). Perhaps the succinic oxidase of E. coli and the lactic oxidase of M. phlei (Brodie, 1959), which do not give phosphorylation in any conditions, are representative of free oxidation. The rate of free oxidation, calculated by Skulachev's formula

$$\frac{P/O_{theor}}{P/O_{meas}} - 1 = \frac{V_{free\ oxid.}}{V_{phosph.\ oxid.}}$$

is, in bacterial preparations oxidizing NAD-bound substrates with P/O about 1, twice as high as that of phosphorylating oxidation, i.e., two-thirds of the substrate is oxidized via a pathway not connected with ATP synthesis (Skulachev, 1962).

We can assume that only one or two (and not all) oxidation–

phosphorylation coupling points function in bacteria (Fujita *et al.,* 1966; Vernon and White, 1957), but at least the most highly organized bacteria can synthesize 3 mole of ATP per mole of oxidized substrate (Asano and Brodie, 1965*b*).

The reason for the peculiar nature of bacterial oxidative phosphorylation may be the very special chemical composition of bacterial membranes (see Chapter I). For instance, *M. lysodeikticus* membranes are totally devoid of nitrogenous phospholipids and sterols, compounds which, according to results obtained in Green's laboratory, play a very important role in the process of oxidative phosphorylation in mitochondria (Green and Lester, 1959; Green, 1962*b*).

The fundamental difference between oxidative phosphorylation in bacteria and that in higher organisms is that this process occurs in a single lipoprotein membrane in bacteria, not in a double membrane, as in mitochondria (see Chapter I). This is obvious in the case of bacteria of the *A. suboxydans* type, where the mesosomes are simply small invaginations of the cytoplasmic membrane (Claus and Roth, 1964), and there are no double membranes at all.

In bacteria with more highly developed mesosomes, such as *M. lysodeikticus,* the membrane preparation which effects oxidative phosphorylation also consists of a mixture of unit membranes (Lukoyanova *et al.,* 1961). Fragments of *A. faecalis* membranes are also of the unit type (Beer, 1960), but mitochondrial fragments can effect oxidative phosphorylation only if the double membrane is preserved. Separation of the mitochondrial membranes leads to the arrest of phosphorylation, which can be restored to some degree by the addition of soluble factors (Green and Hatefi, 1961; Green, 1962). It has been suggested that the second membrane of mitochondria is responsible for the coupling of oxidation and phosphorylation. Weakly coupled oxidative phosphorylation evidently occurs in the unit cytoplasmic membrane of bacteria and the formation of mesosomes may be a means of providing better coupling and, hence, greater efficiency of the process.

In fact, if the respiratory enzymes are localized in the cytoplasmic membrane or in structures anatomically connected with it, the enzymes of phosphorylation must be attached directly to the cytoplasm. Possible evidence of this connection is the release of the oxidative phosphorylation intermediate NAD ∼ E from *A. faecalis* membranes immediately after its formation (Pinchot and Hormanski,

1962*a, b*). In this case the weak coupling of oxidation and phosphorylation and the absence of respiratory control are obviously the result of the weakness and instability of the connection of the respiratory chain with phosphorylation factors.

It would seem that the release of phosphorylation factors in the form of a NAD $\sim$ E type compound from the membranes would lead to rapid cessation of the function of the respiratory chain. This is probably how respiratory control is effected in mitochondria, but we think that the situation in the bacterial cell is rather different. We think that the release of some of the electron carriers from the membrane during oxidative phosphorylation does not lead to malfunction of the respiratory chain, since the links of the chain which are implicated in the formation of the energy-rich compound are present in excess in the membrane and the respiratory chain is not a rigid complex, but is more like a network, via which the electrons continue to move from substrate to oxygen, despite the absence of some links. White (1962, 1965*a*) developed these ideas on the structure of the bacterial respiratory chain from data for *H. parainfluenzae,* but we think that these ideas can be extended to all bacteria.

Mitchell (1961*b,* 1963) has a very original hypothesis as to the mechanism of oxidative and photosynthetic phosphorylation in bacteria, mitochondria, and chloroplasts. The essence of this hypothesis in its simplest form is as follows. The electron-transport chain lies across the lipoprotein membrane, i.e., the substrate approaches it from one side and oxygen from the other. Hence, H+ accumulates on one side of the membrane and OH− on the other, since the membrane is assumed impermeable to ions (Fig. 22).

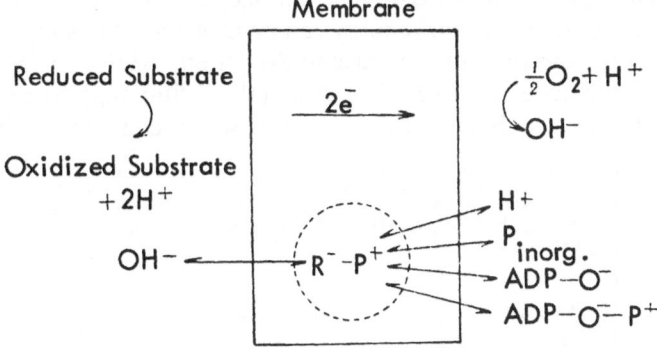

Fig. 22. Mechanism of oxidative phosphorylation, according to Mitchell, 1961.

The membrane contains ATPase, the active center of which is not reached by water as such; OH⁻ ions approach only from the left, and H⁺ only from the right. An excess of OH⁻ on the left intensifies the hydrolysis of ATP, since OH⁻ reacts with the intermediate compound R–P. When there is a deficiency of OH⁻ on the left of the membrane, ATP is synthesized. Since active respiration greatly increases the deficiency of OH⁻ on the left of the membrane, oxidative phosphorylation proceeds. When the selective permeability to ions is destroyed by 2,4-DNP or Dicumarol, for instance, ATP synthesis is arrested, and oxidation and phosphorylation are uncoupled (Mitchell, 1961a).

This hypothesis links oxidative phosphorylation with the transport of various ions through the membrane and with such mechanical transformations of the membrane as compression and swelling. This hypothesis does not require the postulation of various intermediate products of phosphorylation and coupling factors, but great emphasis is laid on the structural integrity of the membranes.

Most investigators of animal and bacterial phosphorylation now base their work on the generally accepted theories of the mechanism of oxidative phosphorylation. It is possible, however, that Mitchell's hypothesis, confirmed, for instance, by the discovery of K⁺- and N⁺-stimulable anisotropic ATPase (Whittam, 1962), which is the foundation stone of the theory of "vectoral metabolism," will be the forerunner of discoveries in many areas of biochemistry. The ideas recently expressed by Brierley and Green (1965) are directly derived from Mitchell's hypothesis. As Fig. 23 shows, Brierley and Green (1965) suggest that mitochondrial membranes have isolated regions in which ATP is synthesized. This endogenous ATP does not leave the site of synthesis, but its energy can be transferred in some way by means of the enzyme mesomerase to exogenous ADP and thus form free ATP. Mitchell and Moyle's ideas (1965) find support in recent investigations which show that there is a direct connection between

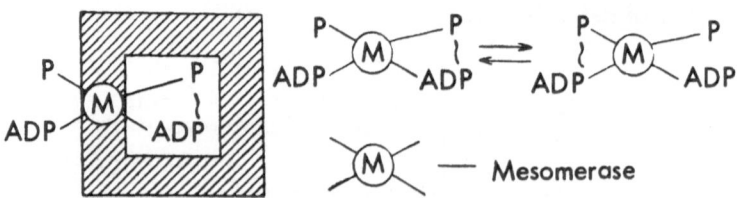

Fig. 23. Schematic representation of location of mesomerase in impermeable region of internal membrane of mitochondrion, according to Brierley and Green, 1965.

proton extrusion and respiration in animal mitochondria (Snoswell, 1966) and between absorption of Ca^{++} and the distribution of H^+ and OH^- outside and inside the mitochondria (Rossi. *et al.,* 1966).

NATURE OF FACTORS COUPLING OXIDATION WITH PHOSPHORYLATION

In dealing with general questions of bacterial oxidative phosphorylation, we discussed the possible structure of phosphorylation factors and represented them as aggregates of enzymes carrying a high-energy bond of varying degrees of complexity. There are two views on the nature of the phosphorylation factors; both are based on data for mitochondrial oxidative phosphorylation. A full account of this question can be found in the review of Bryla and Gardas (1965). Green and his colleagues think that for each phosphorylation point there is one enzyme factor, which they suggest should be called ATP synthetase. This would appear to be supported by the information for *A. faecalis* and the proposed scheme of oxidative phosphorylation (Pinchot, 1965). Yet Pinchot himself is inclined to share the opinion of Racker and Conover (1963), that there are several phosphorylation factors for each coupling point. Pinchot (1965) is inclined towards this view by the fact that a specific tetranucleotide is absolutely essential for oxidative phosphorylation in *A. faecalis,* and the recently discovered requirement of other protein factors, in addition to that mentioned in Fig. 20. A multiplicity of factors is also indicated by the results of investigation of *M. lysodeikticus,* where two protein factors (a heat-labile and a heat-stable one) were discovered. The heat-labile factor lost its coupling power on storage, but still stimulated the ADP–ATP exchange reaction. This indicates the internal complexity of the labile factor itself (Ishikawa and Lehninger, 1962; Yamashita and Ishikawa, 1965).

Not much is known regarding the chemical nature of coupling factors. Only the *A. faecalis* tetranucleotide, consisting of uridylic, adenylic, and guanylic acids in ratio 1:2:1, has been fully determined (Shibko and Pinchot, 1961*b*). The other factors have not been sufficiently purified but, to judge from the available information, the sensitivity of some factors to proteinases, for instance, they are proteins. Watanabe *et al.* (1955) recently reported that *M. phlei* contained a nonprotein phosphorylation factor, the only additional characteristic of which was the presence of absorption maxima at 260 and 340 mμ. Ota (1965) extracted from *E. coli* three protein coupling factors, of which F_1 and F_2 had, in addition to the usual absorption maximum at $\lambda = 280$ mμ, another maximum at $\lambda = 410$ mμ, and F_4 showed three absorption peaks at $\lambda = 352, 371$, and 393 mμ.

The problem of purifying the coupling factors of mitochondria, however, is almost resolved. For instance, Wadkins and Lehninger (1963a) extracted an ADP–ATP exchange enzyme which was identical with the factor specific for the third coupling point and consisted of a protein with a molecular weight of about 35,000.

Factor F_1 (Racker and Conover, 1963), which exhibits ATPase activity and restores coupling of oxidation and phosphorylation at all three points (Schatz and Racker, 1966), has been greatly purified. Intensive research is also being conducted to extract the factor coupling oxidation and phosphorylation between ubiquinone and cytochrome c (Beyer, 1964).

SPECIFICITY OF FACTORS

At present the question of the specificity of the factors cannot be considered resolved in the case of either animal mitochondria or bacterial preparations. This question is particularly complicated in the case of bacteria, owing to the lack of adequate information on coupling points. Brodie (1959) discovered and investigated the interchangeability of particle fractions and factors from *A. vinelandii, M. phlei* and *Corynebacterium creatinovorum.* This universality of bacterial factors for different species of bacteria is of immense interest, but the fact of universality must be treated with caution. For instance, the polynucleotide from *A. faecalis* is required as a coupling factor (in addition to the protein factor) only for *A. faecalis* (Pinchot, 1957a, b), but the protein factor from *M. lysodeikticus* does not stimulate respiration at all, as distinct from the factor from *M. phlei* and *Thiobacillus* X, and is highly active in the ADP–ATP exchange reaction, as distinct from the factor from *A. faecalis* (Ishikawa and Lehninger, 1962).

CONNECTION OF COUPLING FACTORS WITH MEMBRANE CARRYING RESPIRATORY ENZYMES

In the section devoted to oxidative phosphorylation in *M. lysodeikticus,* we tried to envisage the connection of the phosphorylation factor with the membrane.

We think that this question can now be solved most easily by the method extensively used by Racker and his colleagues (Racker, 1963, 1964; Racker and Conover, 1963; Racker and Monroy, 1964). This method consists in partial, successive, reversible destruction of the membrane with various chemical and physical agents (trypsin, urea, NH_4OH, ultrasound). Given the mechanism of action of the agents and analyzing the changes they produce in the action of oxidative phosphorylation enzymes, one can obtain an approximate idea of the role of different types of connection or separate groupings of atoms in the maintenance and accomplishment of the investigated process. Unfortunately, in this case we have only a little information at our disposal and can do no more than conjecture. We think that in *M. lysodeikticus* membranes the phosphorylation factor or factors are held by hydrogen bonds (Ostrovskii and Gel'man, 1965). Magnesium ions neutralize the anionic groups, thus assisting the approach of the components, but do not cross-link them to one another by "bridge bonds." In the intact bacterial cell, polyamines of the spermine type or other substances as yet unknown, may play the role of Mg^{++}. The interaction of mitochondrial coupling factors with the membrane, to judge from the works of Green *et al.* (1963) and Racker and Conover (1963), is similar.

Ultrasonic mitochondrial fragments, like bacterial particles, dissociate in the absence of magnesium ions into a coupling factor and nonphosphorylating particles. A similar process occurs if mitochondria are incubated in a weakly alkaline medium, which probably increases the mutual repulsion of anionic groups.

Like bacterial preparations, mitochondrial particles lose the phosphorylation factor when the hydrogen bonds are destroyed by urea and can again become capable of oxidative phosphorylation when the factor is added.

Thus, the nature of the connection of the coupling factors with the membranes seems very similar in mitochondria and bacteria, but our knowledge in this area is fairly limited. For instance, Wadkins and Lehninger (1963*b*) report that mitochondrial fragments obtained by treatment of the mitochondria with digitonin perform oxidative phosphorylation excellently in the absence of magnesium ions and can be suspended simply in distilled water without suffering any harm.

Racker and Conover (1963) frequently use treatment of mitochondria with ultrasound in a 2% Asolectin solution to separate the phosphorylation factors without injury to the electron-transport chain. Although Asolectin is a mixture of phosphatides, detergents, in fact, other detergents cannot replace it. These effects have not yet been satisfactorily explained, but they are certainly of very great interest for the understanding of the process of oxidative phosphorylation.

Summary

The diverse bacterial kingdom includes groups which differ from one another in the nature of their metabolism. This is clearly brought out by a comparison of the relationship of different groups of bacteria to oxygen. Obligate and facultative anaerobic fermentation bacteria lack cytochromes and use dehydrogenases for electron (hydrogen) transfer in a substrate–substrate chain. Yet there are anaerobes which have acquired cytochromes, but lack terminal oxidases and cannot reduce oxygen. Finally, obligate aerobes and many facultative anaerobes have a complete respiratory chain. Thus, among bacteria we find several groups which differ in their relationship to oxygen. Yet an examination of the ultrastructure of bacteria reveals no correlation between the structure of bacteria and their degree of aerobiosis, since membranous structures are a feature of both aerobes and anaerobes.

As recent investigations have shown, bacteria have membranous structures of varying complexity—from the cytoplasmic membrane alone and simple invaginations of it to multilayered membranous structures (mesosomes). The main element of all bacterial membranous structures is a liproprotein membrane 75–80 Å thick, packed in divers modes. All bacterial membranous structures lack any membrane separating them from the cytoplasm. The diverse forms assumed by bacterial membranous structures indicate their invariable, evolutionally unstabilized nature. Bacterial membranous structures are not independent; they are formed by invagination of the cytoplasmic membrane and remain anatomically connected to it. The special nature of the structure of bacterial membranous structures is responsible for the difficulty of isolating them from the cell, since isolation usually alters the shape and structure and causes their disintegration into separate membranes.

There is weighty evidence in support of a functional analogy between bacterial membranous structures and mitochondria. As distinct from mitochondria, however, bacterial membranous structures contain not only the respiratory chain, but also several other enzyme systems specific to bacteria. Bacterial membranous structures are polyfunctional and are probably an essential device whereby the poorly differentiated bacterial cell can assemble enzyme systems, which require spatial organization and a supply of energy for their activity.

Since the level of respiratory activity of bacteria varies with ambient conditions, it is not surprising that this is reflected in a change in the degree of development of specific membranous structures. The cytoplasmic membrane is of uniform thickness, and the total amount of membranous material in the cell can be increased only by an increase in the extent of the membrane. Such an increase in the surface of the membranous structures in the bacterial cell is associated not only with an increase in respiration rate, but also with certain other, specifically bacterial, functions (Cohen-Bazire and Kunizawa, 1963; Murray, 1963; Stanier, 1964).

Comparison of the fine structure of bacterial membranous structures and mitochondria shows that even the most complex bacterial-membrane systems are very unlike mitochondria. They have no common structural plan and lack double membranes, such as the cristae and outer membrane of the mitochondrion. Yet these biological structures have much in common, since their basic element is a unit lipoprotein membrane 75–80 Å thick. Although the crista is a double membrane, it can be broken down to a "unit" membrane, which can effect electron transport and retains all the components of the respiratory chain (Green, 1962a). In bacterial cells the unit lipoprotein membrane, the carrier of the respiratory-chain enzymes, is only slightly more complicated in the membranous structures, whereas in mitochondria double membranes are formed, and the organelle has a regular shape and structure. It is obvious that the unit lipoprotein membrane in the form found in bacteria and blue-green algae represents the lowest level of spatial organization required for electron transport.

The bacterial membrane also differs from mitochondrial membranes at the level of molecular organization. Analysis of the lipid composition shows that bacterial membranes have no phosphatidylcholine and that the phosphatidylethanolamine:phosphatidylglycerol:diphosphatidylglycerol ratio is very variable. Bacterial membranes have no sterols, but they do often contain carotenoids, which can replace sterols. A significant difference in the composition of bacterial membranes is the absence of di- and polyunsaturated fatty acids, the role of which in the construction of the membrane is fulfilled by branched and cyclic acids. The lipoprotein membranes present in all bacteria do not require all the lipid components present in mitochondrial membranes. Hence, it follows that the most primitive unit lipoprotein membranes could probably develop when a certain protein–lipid ratio was established, provided that hydrophylic and hydrophobic residues, and acid and basic groups, were present in the necessary proportions in the proteins and lipids, and that the branches or rings in the fatty acid residues were arranged in a suitable manner.

It is difficult at present to understand how lipid composition is related to a particular morphological structure, why the lipid composition of mitochondria is much more complicated, and why the ratio of the components varies only in a narrow range. Despite the differences in the lipid composition of bacterial membranes and mitochondria, the construction of bacterial membranes is based on the same intermolecular interactions, although the relative importance of different types of bonds may be different.

Like mitochondrial membranes, bacterial membranes are asymmetric and have subunits whose function is not yet clear on one side. The discovery of subunits on the cytoplasmic membrane and internal membranes of all the investigated species of bacteria indicates the functional similarity of bacterial and mitochondrial membranes, irrespective of whether the subunits contain electron-transport enzymes or oxidative-phosphorylation factors.

The mitochondrion is a structure built according to a strict plan. It has a system of internal membranes and is separated from the cytoplasm by an outer limiting membrane. This gives the mitochondrion a certain autonomy in the cell. The access of metabolites and coenzymes from the cytoplasm to the mitochondrion is obstructed, and special regulatory mechanisms are required to effect exchange between the mitochondrion and the cytoplasm. The mitochondrial matrix and the intercristal spaces contain firmly bound nucleotides (NAD, ADP, ATP), which provide for the operation of the respiratory chain and regulate oxidative phosphorylation.

The absence of mitochondria, the localization of the respiratory chain in a membrane system which has no limiting membrane and lies "exposed" in the cytoplasm and the variation in the shape and size of the membranous structures give rise to several distinctive features in the organization and operation of the respiratory apparatus in bacteria. The absence of enclosed organelles in bacteria means that metabolites and NADH interact directly with the respiratory apparatus. In bacteria there is no spatial separation of the products of anaerobic catabolism in the cytoplasm and the respiratory chain, as is the case in cells with mitochondria. The oxidation of these metabolites is not controlled by the ATP–ADP and NAD–NADH ratios between the membranous structure and the cytoplasm or in the membranous structure itself, since these physiologically important compounds are uniformly distributed throughout the cell owing to the ease of communication between the membranous structures and the cytoplasm. This may be why no respiration-regulating mechanism is required. The availability of substrates for oxidation in the respiratory chain may also account for the high respiration rate in the bacterial cell.

As distinct from the mitochondrial respiratory chain with its constant qualitative and quantitative set of components, bacterial cells have very labile respiratory chains. They can react quickly to changes in the environment and adapt themselves by synthesizing new electron-transferring components. The lability and flexibility of the respiratory apparatus can be regarded as a sign of the evolutionary primitiveness of this group of organisms. This flexibility of the bacterial respiratory apparatus is balanced to some extent by the absence of fine respiration-regulating mechanisms, such as respiratory control or hormonal regulation.

There is no bacterium which has exactly the same set of components as the mitochondrion. It is curious that even within the same species different strains may have different respiratory chains. The set of components different from that in the mitochondrial respiratory chain accounts for the different effect of mitochondrial inhibitors on bacterial chains. Bacteria usually lack the antimycin A-sensitive factor. The reaction of bacteria to inhibitors varies with the phase of development and cultivation conditions.

As in mitochondrial respiratory chains, any substrate of the bacterial chain can reduce all the components of the chain. This indicates that there are links between different electron-transport chains. Examination of the enzyme composition of respiratory chains as a whole reveals features common to all groups of bacteria. Summing up the information on dehydrogenases we noted that there are many dehydrogenases tightly bound to the respiratory chain and that for the dehydrogenation of one substrate there are two enzymes: a firmly bound dehydrogenase not requiring NAD for its action, and a soluble NAD-requiring dehydrogenase.

Bacteria contain a great diversity of ubiquinones and naphthoquinones. it is still not clear why gram-positive bacteria have only naphthoquinones, while gram-negative bacteria have naphthoquinones and ubiquinones. The localization of ubiquinones and naphthoquinones in electron-transport particles and numerous experiments on the reconstruction of oxidation suggest that these components are implicated in the bacterial respiratory chains.

Bacterial cytochromes are also distinguished by their great variety. Many bacteria have several terminal oxidases and the cultivation conditions stimulate the biosynthesis of one or other component. The qualitative and quantitative composition of bacterial cytochromes can change extremely rapidly in response to external factors. Most cytochromes are tightly bound with the membrane fraction and special procedures are required to bring them into solution. Dehydrogenase–cytochrome complexes, respiratory chain fragments, are obtained when bacterial membranes are destroyed.

As in mitochondrial membranes, the lipids of bacterial membranes, including phospholipids, are responsible for the assembly of the respiratory chain components on the membrane. Different respiratory enzymes are specific in their requirement for intact lipid components. It has been suggested that lipids are responsible for the conformation of some electron transport enzymes.

Bacterial respiration, like that of higher organisms, is coupled with the storage of energy in the form of energy-rich compounds. Oxidative phosphorylation, which, as already mentioned, occurs in the mitochondria of higher organisms, has special features in bacterial cells.

There is a lack of convincing evidence to indicate conclusively the localization of oxidative phosphorylation in bacteria. There is no doubt, however, that this process occurs in the lipoprotein membranes and shows some requirement for soluble, mainly protein factors, through which it is connected with metabolism in the cytoplasm. The most distinct feature of oxidative phosphorylation in bacteria is its variability. Its efficiency, expressed as the ratio P/O, varies in a very wide range, depending on the cultivation conditions, phase of growth, and species of bacteria. In the vast majority of cases, the ratio P/O does not exceed unity for oxidation of NAD-bound substrates, which is only one-third of the corresponding value for mitochondria. In addition, specific inhibitors of oxidative phosphorylation in mitochondria (DNP, oligomycin, etc.) usually have no effect at all, or a much weaker effect, on bacterial preparations. In certain bacteria, however, oxidative phosphorylation is just as efficient and sensitive to inhibitors as in mitochondria. Oxidative phosphorylation in subcellular bacterial systems usually requires soluble factors of various kinds. The list of factors at present includes heat-labile and heat-stable proteins and polynucleotides.

The discussed features of oxidative phosphorylation in bacteria suggest that individual stages of this process are different in different bacteria, and that a common feature in bacteria is the ability to use nonphosphorylated intermediate energy-rich compounds. This is one of the reasons for the low P/O values. In view of what has been said, it is difficult to characterize the mechanism of bacterial oxidative phosphorylation as a whole. However, the mechanism of the process is fully clarified for at least the first point of coupling of respiration with phosphorylation in *A. faecalis*. It consists essentially in the formation of an energy-rich compound of the oxidized electron carrier and the protein factor, to which inorganic phosphate is then attached. The phosphate and energy are then transferred to ADP. The formation of the carrier $\sim$ factor compound is accompanied by its release from the membrane but, strange as it seems, bacterial preparations show hardly any regulation of respiration, which is so characteristic of mitochondria. The reason for the special nature of bacterial oxidative phosphorylation, and electron transport, lies in the localization of these processes in a unit lipoprotein membrane, which is open to interaction with the numerous processes in the cytoplasm. This inevitably leads to simplification of regulatory mechanisms.

Summing up the account of the structure and function of the bacterial respiratory apparatus, we must acknowledge its fairly primitive nature, despite the presence of many features in common with highly organized cells (Palade, 1964). Hence, present-day bacteria provide interesting material from the viewpoint of investigation of the evolution of structure and function of the respiratory apparatus down the evolutionary ladder—to the predecessors of bacteria, as well as up the ladder—to organisms with mitochondria.

We must mention that there is no general agreement on the evolutionary position of bacteria. While Gale (1962) regards bacteria as the most ancient organisms, Kerkut (1960) thinks that it is difficult at present to choose between the following theories: 1) bacteria gave rise to protozoa; 2) bacteria are degraded protozoa; 3) bacteria and protozoa are independent groups; 4) bacteria form an inhomogeneous group which includes truly primitive organisms and secondarily simplified forms. Stanier and Van Niel put bacteria and bluegreen algae into a special group, the prokaryotes, as distinct from the animal and plant kingdom, the eukaryotes (Stanier and Van Niel, 1962; Stanier, 1964). At any rate, bacteria are definitely among the most simply organized living creatures and are most like our conception of protobionts—the first inhabitants of the earth.

According to Oparin, at the dawn of life the most simply organized systems were coacervates. They effected chemical reactions, accumulated substances from the external medium, and excreted reaction products into it (Oparin, 1962, 1966; Oparin and Serebrovskaya, 1963). The structural element of the most primitive coacervate systems was probably a superficial lipoprotein membrane. In the course of a long evolution, coacervate systems under the action of selection could have given rise to the first living cells. The first organisms were anaerobes and heterotrophs. These were followed by photosynthetic forms and finally by autotrophic and heterotrophic aerobic forms (Oparin, 1957, 1960, 1962).

Thus, the process of respiration itself in the modern sense appeared relatively late in evolution. How might it have originated?

In the preaerobic epoch, electron transport was effected by a shortened respiratory chain and involved the reduction of inorganic electron acceptors ("nitrate" and "sulfate" respiration). Organisms having such a respiratory apparatus have survived till the present day. They have a respiratory apparatus, but in a number of cases lack cytochrome oxidase. Yet many of them can switch from anaerobic to aerobic metabolism. The next evolutionary step based on anaerobic forms was the appearance of photosynthetic bacteria, the structural manifestation of which was the appearance of chromatophores anatomically connected with the cytoplasmic membrane. Finally, the appearance of oxygen led to the formation of the respiratory apparatus and of cytochrome oxidase, which interacts with oxygen. In the earliest inhabitants of the earth, as in present-day bacteria, the lipoprotein membrane played an exceptional role as

the main factor in structure formation and isolation of the protobionts from the surrounding medium. Mentally comparing protobionts with bacteria, we realize that the surface location of the respiratory enzymes and the imperfections of oxidative phosphorylation are something theoretically natural for bacteria.

Tracing the evolution of the cytoplasmic membrane and membrane systems of present-day bacteria to mitochondria is an infinitely
more difficult task, since the factual material is immense and its
interpretation abounds in contradictions. A fascinating hypothesis
in connection with the structure of the bacterial respiratory apparatus
is that of Robertson (1959), who postulates that all membrane
systems of modern cells, including mitochondria, are interconnected
and are derived from a common plasma membrane. This hypothesis
finds some confirmation in the fact that in highly developed organisms the function of electron transport has not been entirely handed
over to the mitochondria. For instance, Emmelot et al., (1964) report
that the cytoplasmic membrane of liver cells contains, in addition to
many other enzymes, NADH:cytochrome c reductase, i.e., a large
portion of the respiratory chain, while Kono and Colowick (1961)
reported even earlier that NADH oxidase is present in muscle-cell
membranes. It was found recently that the cells of some organisms
contain so-called microbodies, which effect free oxidation of different
substrates (Baudhuin et al., 1965). Robertson's view was developed
by several authors (Moore and McAlear, 1961; Vinnikov, 1964; Wildon and Mercer, 1963). Although the hypothesis of the morpho-
genetic role of the cytoplasmic membrane in the origination of mitochondria is very attractive, it is not the only one. A number of other
hypotheses include the idea that the biogenesis of mitochondria may
be connected with the nuclear membrane or even with the membranes of the endoplasmic reticulum. The fact is that the membrane
of the endoplasmic reticulum contains a special set of respiratory
enzymes (Novikoff, 1962):

$$\text{Substrate} \rightarrow \text{NAD} \rightarrow \text{Flavoprotein} \rightarrow \text{Cytochrome } b_5 \rightarrow \text{Cytochrome } c \rightarrow ?$$

The oxidative phosphorylation in the nuclei of higher organisms
may be a reflection of mechanisms which operated at one time (Pen-
nial et al., 1962, 1964; Gaitskhoki, 1964; Allfrey and Mirsky, 1958;
McEwen et al., 1963). This may prove to be more than a fleeting
hypothesis in view of the ideas developed by Scheide et al., (1964),

who suggest that the mitochondrion develops in ontogenesis from the nuclear membrane, thus indicating the antiquity of nuclear oxidative phosphorylation.

A special hypothesis regarding the evolution of the cell is that which postulates that the mitochondrion is a product of the symbiosis of very ancient bacteria and certain unicellular organisms and has developed in recent time (Ris and Plaut, 1962). This idea is confirmed by the discovery of DNA and ribosomes in mitochondria and chloroplasts (Nass *et al.,* 1965; Sisakyan and Bekina, 1964). Despite its rather fantastic nature, Ris's theory suggests a direct pathway of evolution from the most primitive bacteria, or their predecessors, to mitochondria.

Literature Cited

Abel, K., de Schmertzing, H., and Peterson, J., 1963. *J. Bacteriol.* 85:1039.
Abram, D., 1965. *J. Bacteriol.* 89:855.
Abrams, A., 1959. *J. Biol. Chem.* 234:383.
Abrams, A., McNamara, P., and Johnson, F., 1960. *J. Biol. Chem.* 235:3659.
Abrams, A., Nielsen, L., and Thaemert, J., 1964. *Biochim. Biophys. Acta* 80:325.
Adams, E., and Newberry, S., 1961. *Biochem. Biophys. Res. Commun.* 6:1.
Adams, J., Painter, B., and Payne, W., 1963. *J. Bacteriol.* 86:548, 558.
Adelson, J., Asano, A., and Brodie, A., 1964. *Proc. Natl. Acad. Sci. U.S.* 51:402.
Ahmad, K., Schneider, H., and Strong, F., 1950. *Arch. Biochem. Biophys.* 28:281.
Akamatsu, Y., and Nojima, S., 1965. *J. Biochem.* 57:430.
Akamatsu, Y., Ono, Y., and Nojima, S., 1966. *J. Biochem.* 59:176.
Aleem, M., 1965. *Biochim. Biophys. Acta* 107:14.
Aleem, M., 1966a. *Biochim. Biophys. Acta* 113:216.
Aleem, M., 1966b. *J. Bacteriol.* 91:729.
Aleem, M., and Lees, H., 1963. *Can. J. Biochem. Physiol.* 41:763.
Aleem, M., Lees, H., and Nicholas, D., 1963. *Nature* 200:759.
Aleem, M., and Nason, A., 1959. *Biochem. Biophys. Res. Commun.* 1:323.
Aleem, M., and Nason, A., 1960. *Proc. Natl. Acad. Sci. U.S.* 46:763.
Alexander, M., 1956. *Bacteriol. Rev.* 20:67.
Allen, M., and Van Niel, C., 1952. *J. Bacteriol.* 64:397.
Allfrey, V., and Mirsky, A., 1958. *Proc. Natl. Acad. Sci. U.S.* 44:981.
Ambe, K., and Crane, F., 1959. *Science* 129:98.
Ambler, R., 1963. *Biochem. J.* 89: 341, 349.
Anderson, J., 1964. *Biochem. J.* 91:8.
Arcos, J., and Argus, M., 1964. *Biochemistry* 3:2028.
Argaman, M., and Razin, S., 1965. *J. Gen. Microbiol.* 38:153.
Arima, K., and Oka, T., 1965a. *J. Bacteriol.* 90:734, 744.
Arima, K., and Oka, T., 1965b. *J. Biochem.* 58:320.
Aronson, A., 1966. *J. Mol. Biol.* 15:505.
Asano, A., Adelson, J., and Brodie, A., 1964. *6th Intern. Congr. Biochem. Abstr. (New York)* X:773.
Asano, A., and Brodie, A., 1963. *Biochem. Biophys. Res. Commun.* 13:416.
Asano, A., and Brodie, A., 1964. *J. Biol. Chem.* 239:4280.
Asano, A., and Brodie, A., 1965a. *Biochem. Biophys. Res. Commun.* 19:121.
Asano, A., and Brodie, A., 1965b. *J. Biol. Chem.* 240:4002.
Asano, A., Brodie, A., Wagner, A., Wittreich, P., and Folkers, K., 1962. *J. Biol. Chem.* 237: PC 2412.
Asano, A., Kaneshiro, T., and Brodie, A., 1965. *J. Biol. Chem.* 240:895.
Asnis, R., Vely, V., and Glick, M., 1956. *J. Bacteriol.* 72:314.
Asselineau, J., 1962. *Les lipides bactériens,* Hermann, Paris.
Asselineau, J., and Lederer, E., 1960. In: Bloch, K. (ed.), *Lipide Metabolism.* Wiley, New York.
Aubert, J., Millet, J., and Milhaud, G., 1959. *Ann. Inst. Pasteur* 96:559.

Avakyan, A. A., Pavlova, I. B., Kats, L. N., and Levina, E. N., 1967. *J. Hygiene, Epidemiol. Microbiol., Immunol.,* Prague (in press).

Azerad, R., Bleiler-Hill, R., and Lederer, E., 1965. *Biochem. Biophys. Res. Commun.* 19:194.

Azoulay, E., 1964. *Biochim. Biophys. Acta* 92:458.

Azoulay, E., and Couchoud-Beaumont, P., 1965. *Biochim. Biophys. Acta* 110:301.

Baker, P., Northcote, D., and Peters, R., 1962. *Nature* 195:661.

Ball, E., and Joel, C., 1962. *Intern. Rev. Cytol.* 13:99.

Baltscheffsky, H., and von Stedingk, L.-V., 1966. *Biochem. Biophys. Res. Commun.* 22: 722.

Bangham, A., and Horne, R., 1964. *J. Mol. Biol.* 8:660.

Barban, P. S., 1963. In: *Rickettsial and Viral Infections, Perm Med. Inst.,* pp. 239, 247.

Barret, J., 1956. *Biochem. J.* 64:626.

Bartsch, R., and Kamen, M., 1958. *J. Biol. Chem.* 230:41.

Bastarrachea, F., and Goldman, D., 1961. *Biochim. Biophys. Acta* 50:174.

Baudhuin, P., Müller, M., Poole, B., and De Duve, C., 1965. *Biochem. Biophys. Res. Commun.* 20:53.

Baum, R., and Dolin, M., 1963. *J. Biol. Chem.* 238: PC 4109.

Baum, R., and Dolin, M., 1965. *J. Biol. Chem.* 240:3425.

Beer, M., 1960. *J. Bacteriol.* 80:659.

Beinert, H., 1960. *The Enzymes,* Vol. 2A, Academic Press, Inc., New York, p. 340.

Beinert, H., Griffiths, D. Wharton, D., and Sands, R., 1962. *J. Biol. Chem.* 237:2337.

Beinert, H., and Lee, W., 1961. *Biochem. Biophys. Res. Commun.* 5:40.

Belitser, V. A., and Tsybakova, E. T., 1939. *Biokhimiya* 4:516.

Benson, A., 1964. *Ann. Rev. Plant Physiol.* 15:1.

Bentley, R., and Schlechter, L., 1960. *J. Bacteriol.* 79:346.

Benziman, M., and Galanter, Y., 1964. *J. Bacteriol.* 88:1011.

Benziman, M., and Perez, L., 1965. *Biochem. Biophys. Res. Commun.* 19:127.

Beyer, R., 1964. *Federation Proc.* 23:432.

Bieber, T., 1964. *Biochem. Biophys. Res. Commun.* 16:501.

Biryuzova, V. I., 1960. *Dokl. Akad. Nauk SSSR* 135:1519.

Biryuzova, V. I., Lukoyanova, M. A., Gel'man, N. S., and Oparin, A. I., 1964. *Dokl. Akad. Nauk SSSR* 156:198.

Biryuzova, V. I., and Meisel', M. N., 1964. In: *Molecular Biology,* Nauka Press, Moscow, p. 215.

Bishop, D., and King, H., 1962. *Biochem. J.* 85:550.

Bishop, D., Pandya, K., and King, H., 1962. *Biochem. J.* 83:606.

Bladen, H., Nylen, M., and Fitzgerald, R., 1964. *J. Bacteriol.* 88:763.

Bladen, H., and Waters, J., 1963. *J. Bacteriol.* 86:1339.

Blair, P., Oda, T., Green, D., and Fernandez-Móran, H., 1963. *Biochemistry* 2:756.

Blakley, E., and Cifferi, O., 1961. *Can. J. Microbiol.* 7:61.

Blaylock, B., and Nason, A., 1963. *J. Biol. Chem.* 238:3453.

Blyumenfel'd, L. A., and Purmal', A. P., 1964. In: *Enzymes,* Nauka Press, Moscow, p. 215.

Boatman, F., 1964. *J. Cell. Biol.* 20:297.

Boatman, F., and Douglas, H., 1961. *J. Biophys. Biochem. Cytol.* 11:469.

Bone, D., 1963. *Biochim. Biophys. Acta* 67:589.

Bone, D., Bernstein, S., and Vishniak, W., 1963. *Biochim. Biophys. Acta* 67:581.

Borisov, A. Yu., and Mokhova, E. N., 1964. *Pribory i Tekhn. Eksperim.* 2:145.

Borst, P., Loos, J., Christ, E., and Slater, E., 1962. *Biochim. Biophys. Acta* 62:509.

Bourne, C., 1958. In: Cook, R. (ed.), *Cholesterol, Chemistry, Biochemistry, and Pathology,* Academic Press, Inc., New York, p. 349.

Bradfield, J., 1956. *Symp. Soc. Gen. Microbiol.* 6:296.

Bragg, P., 1965a. *Biochim. Biophys. Acta* 96:263.

Bragg, P., 1965b. *J. Bacteriol.* 90:1498.

Brenner, S., and Horne, R., 1959. *Biochim. Biophys. Acta* 34:103.

Bresler, S. E., 1963. *An Introduction to Molecular Biology,* USSR Academy of Sciences Press, Moscow and Leningrad. To be published by Gordon and Breach Science Publishers, Inc., New York.

Brieger, E., 1963. *Structure and Ultrastructure of Microorganisms,* Academic Press, Inc., New York.

Brierley, G., and Green, D., 1965. *Proc. Natl. Acad. Sci. U.S.* 53:73.

Brierley, G., and Merola, A., 1962. *Biochim. Biophys. Acta* 64:205.

Brodie, A., 1959. *J. Biol. Chem.* 234:398.

Brodie, A., 1961a. In: Wolstenholme, G., and O'Connor, C. (eds.), *Quinones in Electron Transport,* Churchill Ltd., London.

Brodie, A., 1961b. *Federation Proc.* 20:995.

Brodie, A., 1963. In: Colowick, S., and Kaplan, N. (eds.), *Methods in Enzymology,* Academic Press, Inc., New York, p. 284.

Brodie, A., and Adelson, J., 1965. *Science* 149:265.

Brodie, A., and Ballantine, J., 1960. *J. Biol. Chem.* 235:232.

Brodie, A., Davis, B., and Fieser, L., 1958. *J. Am. Chem. Soc.* 80:6454.

Brodie, A., and Gray, C., 1955. *Biochim. Biophys. Acta* 17:146.

Brodie, A., and Gray, C., 1956a. *Biochim. Biophys. Acta* 19:384.

Brodie, A., and Gray, C., 1956b. *J. Biol. Chem.* 219:853.

Brodie, A., and Gray, C., 1957. *Science* 125:534.

Brodie, A., Weber, M., and Gray, C., 1957. *Biochim. Biophys. Acta* 25:448.

Bronk, I., 1963. *Proc. Natl. Acad. Sci. U.S.,* 50:524.

Brown, A., 1962a. *Biochim. Biophys. Acta* 62:132.

Brown, A., 1962b. *Biochim. Biophys. Acta* 58:514.

Brown, A., 1964a. *Bacteriol. Rev.* 28:296.

Brown, A., 1964b. *Biochim. Biophys. Acta* 93:136.

Brown, A., 1965. *J. Mol. Biol.* 12:491.

Brown, A., Drummond, D., and North, R., 1962. *Biochim. Biophys. Acta* 58:514.

Brown, A., Shorey, C., and Turner, H., 1965. *J. Gen. Microbiol.* 41:225.

Brown, J., *1961. Biochim. Biophys. Acta* 52:368.

Brown, J., 1965. *Biochim. Biophys. Acta* 94:97.

Brown, J. and Cosenza, B., 1964. *Nature* 204:802.

Bruemmer, J., Wilson, P., Glenn, J., and Crane, F., 1957. *J. Bacteriol.* 73:113.

Brya, J., and Gardas, A., 1965. *Postepy Biochim.* 11:395.

Bulen, W., Burns, R., and Le Comte, J., 1964. *Biochem. Biophys. Res. Commun.* 17:265.

Bultow, D., and Levedahl, B., 1964. *Ann. Rev. Microbiol.* 18:167.

Burgos, J., and Redfearn, E., 1965. *Biochim. Biophys. Acta* 110:475.

Burrous, S., and Wood, W., 1962. *J. Bacteriol.* 84:365.

Burton, D., and Glover, J., 1965. *Biochem. J.* 94:27P.

Butler, J., Crathorn, A., and Hunter, G., 1958. *Biochem. J.* 69:544.

Butler, J., Godson, G., and Hunter, G., 1961. In: Harris, R. (ed.), *Protein Biosynthesis,* Academic Press, Inc., New York.

Butt, W., and Lees, H., 1958. *Nature* 182:732.

Campbell, N., and Aleem, M., 1965. *Antonie van Leeuwenhoek, J. Microbiol. Serol.* 31: 124, 137.

Campbell, J., Hogg, L., and Strasdine, G., 1962. *J. Bacteriol.* 83:1155.

Carr, N., 1964. *Biochem. J.* 91:28P.

Carr, N., and Exell, G., 1965. *Biochem. J.* 96:688.

Castor, L., and Chance, B., 1955. *J. Biol. Chem.* 217:453.

Castor, L., and Chance, B., 1959. *J. Biol. Chem.* 234:1587.

Cerletti. P., Strom, R., and Giordano, M., 1965. *Biochem. Biophys. Res. Commun.* 18:259.

Chaloupka, J., and Vereš, K., 1961. *Z. Allgem. Mikrobiol.* 1:325.

Chance, B., 1951. *Rev. Sci. Inst.* 22:619.

Chance, B., 1952. *Nature* 169:215.

Chance, B., 1964. *6th Intern. Congr. Biochem. Abstr. (New York)* VIII:617.

Chance, B., 1965. *J. Gen. Physiol.* 49(1) part 2:163.

Chance, B., and Hagihara, B., 1962. In: *Intracellular Respiration, Phosphorylating and Nonphosphorylating Oxidation Reactions* (Russian edition), Proceedings of the Fifth International Biochemical Congress, Symposium V, USSR Academy of Sciences Press, Moscow, p. 10. Available in English as: *International Congress of Biochemistry, 5th, Moscow, 1961. Biochemistry: Proceedings,* 9 vols., Sissakian, N. M. (ed.), Pergamon Press, London.

Chance, B., Horio, T., Kamen, M., and Taniguchi, S., 1966. *Biochim. Biophys. Acta* 112:1.

Chance, B., and Parsons, D., 1963. *Science* 142:1176.

Chance, B., and Redfearn, E., 1961. *Biochem. J.* 80:632.

Chance, B., and Smith, L., 1955. *Nature* 175:843.

Chance, B., and Williams, G., 1955. *J. Biol. Chem.* 217:409.

Chance, B., and Williams, G., 1956. *Advan. Enzymol.* 17:65.

Chang, J., and Lascelles, J., 1963. *Biochem. J.* 89:503.

Chapman, G., 1959. *J. Bacteriol.* 78:96.

Chapman, G., and Kroll, A., 1957. *J. Bacteriol.* 73:63.

Chappell, J., and Greville, G., 1963. *Biochem. Soc. Symp.* 23:39.

Chen, S., 1964. *Nature* 202:1135.

Chin, C., 1952. *2e Congr. Intern. Biochim. Paris. Résumés Communs,* p. 277.

Cho, K., and Salton, M., 1964. *Biochim. Biophys. Acta* 84:773.

Cho, K., and Salton, M., 1966. *Biochim. Biophys. Acta* 116:73.

Classification and Nomenclature of Enzymes (Russian translation), Foreign Literature Press, Moscow, 1962.

Claus, G., and Roth, L., 1964. *J. Cell. Biol.* 20:217.

Clayton, R., 1955. *Arch. Mikrobiol.* 22:195.

Cohen-Bazire, G., and Kunizawa, R., 1963. *J. Cell. Biol.* 16:401.

Cohen-Bazire, G., Kunizawa, R., and Poindexter, J., 1966. *J. Gen. Microbiol.* 42:301.

Çohen-Bazire, G., Pfennig, N., and Kunizawa, R., 1964. *J. Cell. Biol.* 22:207.

Cohen, M., and Wainio, W., 1963. *J. Biol. Chem.* 238:879.

Cohn, D., 1956. *J. Biol. Chem.* 221:413.

Cohn, D., 1958. *J. Biol. Chem.* 233:299.

Cole, R., 1965. *Bacteriol. Rev.* 29:326.

Conover, T., and Ernster, L., 1963. *Biochim. Biophys. Acta* 67:268.

Conti, S., and Gettner, M., 1962. *J. Bacteriol.* 83:544.

Conti, S., and Hirsch, P., 1965. *J. Bacteriol.* 89:503.

Cook, T., 1964. *Dissertation Abstr.* 24:2653.

Cook, W., and Martin, W., 1962. *Biochim. Biophys. Acta* 56:362.

Cota-Robles, E., and Coffman, M., 1964. *J. Ultrastruct. Res.* 10:304.

Cota-Robles, E., Marr, A., and Nilson, E., 1958. *J. Bacteriol.* 75:243.

Coval, M., Horio, T., and Kamen, M., 1961. *Biochim. Biophys. Acta* 51:246.

Criddle, R., Bock, R., Green, D., and Tisdale, H., 1962. *Biochemistry* 1:827.

Cunningham, W., Hall, C., and Crane, F., 1964. *6th Intern. Congr. Biochem. Abstr. (New York)* VIII:647.

Cunningham, W., Hall, C., Crane, F., and Das, M., 1965. *Federation Proc.* 24(2) part 1:296.

Dalgarno, L., and Birt, L., 1963. *Biochem. J.* 87:586.

Dallam, R., 1961. *Biochem. Biophys. Res. Commun.* 4:106.

Danielson, L., and Ernster, L., 1962. *Nature* 194:155.

Dart, P., and Mercer, F., 1963*a*. *Arch. Mikrobiol.* 46:382.

Dart, P., and Mercer, F., 1963*b*. *Arch. Mikrobiol.* 47:1.

Das, M., and Crane, F., 1964. *Biochemistry* 3:696.

Das, M., Myers, D., and Crane, F., 1964. *Biochim. Biophys. Acta* 84:618.

Daubner, I., 1962. *Spisy Prirodovedecke Fak. Univ. Brně* 10:472.

Davson, H., and Danielli, J., 1943. *The Permeability of Natural Membranes*, Cambridge University Press, New York.

Dawes, E., and Ribbons, D., 1964. *Bacteriol. Rev.* 28:126.

DeBoer, W., La Rivière, J., and Houwink, A., 1961. *Antonie van Leeuwenhoek J. Microbiol. Serol.* 27:444.

Deborin, G. A., Baranova, V. Z., and Zhukova, I. G., 1966. *Dokl. Akad. Nauk SSSR* 166:231.

Deeb, S., and Hager, L., 1964. *J. Biol. Chem.* 239:1024.

De Ley, J., 1960. *Nature* 188:331; *J. Appl. Bacteriol.* 23:400.

De Ley, J., 1962. *Biochem. J.* 84:9P.

De Ley, J., 1963. *Biochem. Soc. Symp.* 23:138.

De Ley, J., 1964. *Ann. Rev. Microbiol.* 18:17.

De Ley, J., and Dochy, R., 1960*a*. *Biochim. Biophys. Acta* 40:277.

De Ley, J., and Dochy, R., 1960*b*. *Biochim. Biophys. Acta* 42:538.

De Ley, J., and Kersters, K., 1964. *Bacteriol. Rev.* 28:164.

De Ley, J., and Schell, J., 1959. *Biochim. Biophys. Acta* 35:154.

De Ley, J., and Schell, J., 1962. *J. Gen. Microbiol.* 29:589.

Delwiche, C., Burge, W., and Malavolta, E., 1961. In: *Abstracts of Sectional Papers of Fifth International Biochemical Congress,* Vol. II (Russian edition), USSR Academy of Sciences Press, Moscow, p. 393. Available in English as: *International Congress of Biochemistry, 5th, Moscow, 1961. Biochemistry: Proceedings,* 9 Vols., Sissakian, N. M. (ed.), Pergamon Press, London.

De Petris, S., Karlsbad, G., and Kessel, R., 1964. *J. Gen. Microbiol.* 35:373.

Dillon, L., 1962. *Evolution* 16:102.

Dixon, M., and Webb, E., 1961. *Enzymes* (Russian translation), Foreign Literature Press, Moscow. Originally published in English by Academic Press, Inc., New York, 1959.

Dixon, R., 1964. *Arch. Mikrobiol.* 48:166.

Doeg, K., 1961. *Federation Proc.* 20:44.

Doi, R., and Halvorson, H., 1961. *J. Bacteriol.* 81:51.

Dolin, M. 1963*a*. In: Gusalus, U., and Stainer, R. (eds.), *The Metalobism of Bacteria* (Russian translation), Foreign Literature Press, Moscow, p. 316. Originally published in English by Academic Press, Inc., New York: Gusalus, U., and Stainer, R. (eds.), *Bacteria: A Treatise on Structure and Function,* 5 Vols., 1960–1964.

Dolin, M., 1963*b*. *Ibid., p. 414.*

Doman, N. G., and Tikhonova, G. V., 1965. *Usp. Sovrem. Biol.* 60:238.

Domermuth, C., Nielsen, M., Freundt, E., and Birch-Andersen, A., 1964*a*. *J. Bacetriol.* 88:727.

Domermuth, C., Nielsen, M., Freundt, E., and Birch-Andersen, A., 1964*b*. *J. Bacteriol.* 88:1428.

Dowben, R., and Koeler, W., 1961. *Arch. Biochem. Biophys.* 93:496.

Downey, R., 1963. *J. Bacteriol.* 84:953.

Downey, R., 1964. *J. Bacteriol.* 88:904.

Downey, R., 1966. *J. Bacteriol.* 91:634.

Downey, R., Georgi, C.,and Militzer, W.,1962. *J. Bacteriol.* 83:1140.

Drapeau, G., and MacLeod, R., 1963. *J. Bacteriol.* 85:1413.
Dugan, P., and Lundgren, D., 1965. *J. Bacteriol.* 89:825.
Dworkin, M., and Niederpruem, D., 1964. *J. Bacteriol.* 87:316.

Eagon, R., and Carson, J., 1965. *Can. J. Microbiol.* 11:193.
Eagon, R., and Williams, A., 1959. *Arch. Biochem. Biophys.* 79:401.
Echlin, P., and De Lamater, E., 1962. *Exptl. Cell Res.* 26:229.
Echlin, P., and Morris, J., 1965. *Biol. Rev.* 40:143.
Edebo, L., 1961. *Acta Pathol. Microbiol. Scand.* 52:372, 384; 53:121.
Edwards, M., and Stevens, R., 1963. *J. Bacteriol.* 86:414.
Edwards, S., and Ball, E., 1954. *J. Biol. Chem.* 209:619.
Egami, F., Ishimoto, M., and Taniguchi, S., 1961. In: Falk, J., Lemberg, R., and Morton,
      R., (eds.), *Hematin Enzymes,* Vol. 2, Pergamon Press, Oxford, p. 392.
Ehrenberg, A., and Kamen, M., 1965. *Biochim. Biophys. Acta* 102:333.
Eiserling, F., and Romig, W., 1962. *J. Ultrastruct. Res.* 6:540.
Ells, A., 1959. *Arch. Biochem. Biophys.* 85:561.
Emmelot, P., Bos, C., Benedett, E., and Rumke, P., 1964. *Biochem. Biophys. Res.
      Commun.* 90:126.
Ermachenko, V. A., Lisenkova, L. L., and Lozinov, A. B., 1966. *Mikrobiologiya* 25:242.
Ernster, L., 1961. In: *IUB/IUBS Symposium on Biological Structure and Function,* Vol. 2,
      Academic Press, Inc., London, p. 139.
Ernster, L., and Lee, Chuan-pu, 1964. *Ann. Rev. Biochem.* 33:729.
Erwin, J., and Bloch, K., 1964. *Science* 143:1006.
Estabrook, R., 1962. *Biochim. Biophys. Acta* 60:236.
Estabrook, R., and Holowinsky, A., 1961. *J. Biophys. Biochem. Cytol.* 9:19.

Falcone, A., Shug, A., and Nicholas, D., 1962. *Biochem. Biophys. Res. Commun.* 9:126.
Falcone, A., Shug, A., and Nicholas, D., 1963. *Biochim. Biophys. Acta* 77:199.
Falk, J., 1964. *Porphyrins and Metalloporphyrins. BBA Library,* Vol. 2, American Elsevier
      Publishing Company, Inc., New York.
Falk, J., Lemberg, R., and Morton, R., (eds.) 1961. *Hematin Enzymes,* Vol. 2, Pergamon
      Press, Oxford, p. 432.
Feldman, W., and O'Kane, D., 1960. *J. Bacteriol.* 80:218.
Fellman, J., and Mills, R., 1960. *J. Bacteriol.* 79:800.
Feo, F., Gabriel, L., and Terranova, T., 1962. *Z. Physiol. Chem.* 329:188.
Fernandez-Móran, H., 1962. *Circulation* 26:1039.
Fernandez-Móran, H., 1963. *Science* 140:380.
Fernandez-Móran, H., Oda, T., Blair, P., and Green, D., 1964. *J. Cell Biol.* 22:63.
Few, A., 1955. *Biochim. Biophys. Acta* 16:137.
Fewson, S., and Nicholas, D., 1961. In: *Abstracts of Sectional Papers of the Fifth Inter-
      national Biochemical Congress* (Russian edition), Vol. I, USSR Academy of
      Sciences Press, Moscow, p. 309. Available in English as: *International Congress
      of Biochemistry, 5th, Moscow, 1961. Biochemistry: Proceedings,* 9 vols.,
      Sissakian, N. M. (ed.), Pergamon Press, London.
Fewson, C., and Nicholas, D., 1961a. *Biochim. Biophys. Acta* 48:210.
Fewson, C., and Nicholas, D., 1961b. *Biochim. Biophys. Acta* 49:335.
Fischer, Y., and Laudelout, H., 1965a. *Biochim. Biophys. Acta* 110:204.
Fischer, Y., and Laudelout, H., 1965b. *Biochim. Biophys. Acta* 110:259.
Fitz-James, P., 1960. *J. Biophys. Biochem. Cytol.* 8:507.
Fitz-James, P., 1962. *J. Bacteriol.* 84:104.
Fitz-James, P., 1964a. *J. Bacteriol.* 87:1477.
Fitz-James, P., 1964b. *J. Bacteriol.* 87:667.
Fitz-James, P., 1964c. *J. Bacteriol.* 87:1483.

Fitz-James, P., 1965. In: *Structure and Function in Microorganisms,* Vol. 15, Soc. Gen. Microbiol., p. 369.

Fleischer, S., 1964. *6th Intern. Congr. Biochem. Abstr. (New York)* VIII:605.

Fleischer, S., Brierley, G., Klouwen, H., and Slautterback, D., 1962. *J. Biol. Chem.* 237: 3264.

Fleischer, S., and Klouwen, H., 1961. *Biochem. Biophys. Res. Commun.* 5:378.

Fleischer, S., Klouwen, H., and Brierley, G., 1961. *J. Biol. Chem.* 236:2936.

Francis, M., Hughes, D., Kornberg, H., and Phizackerley, P., 1963. *Biochem. J.* 89:430.

Francis, M., and Phizackerley, P., 1965. *Biochem. J.* 95:25P.

Freimer, E., 1963. *J. Exptl. Med.* 117:377.

Freimer, E., and Krause, R., 1960. *Federation Proc.* 19:244.

Frieden, E., 1964. In: Kasha, M., and Pullman, B. (eds.), *Horizons in Biochemistry* (Russian translation), Mir Press, Moscow, p. 360. Originally published in English by Academic Press, Inc., New York, 1962.

Frieden, E., Osaka, S., and Kobayashi, H., 1965. *J. Gen. Physiol.* 49(1) part 2:213.

Fuchs, G., 1965. *Arch. Mikrobiol.* 50:25.

Fujita, F., and Sato, R., 1963. *J. Biochem.* 77:690.

Fujita, M., Ishikawa, S., and Shinazono, N., 1966. *J. Biochem. (Japan)* 59:104.

Fujita, T., Itagaki, E., and Sato, R., 1963. *J. Biochem.* 53:282.

Fulco, A., Levy, R., and Bloch, K., 1964. *J. Biol. Chem.* 239:998.

Gaffron, H., 1964. In: Kasha, M., and Pullman, B. (eds.), *Horizons in Biochemistry* (Russian translation), Mir Press, Moscow, p. 49. Originally published in English by Academic Press, Inc., New York, 1962.

Gaitskhoki, V. S., 1964. *Usp. Sovrem. Biol.* 57:30.

Gale, E., 1959. *Proc. Roy. Soc. (London), Ser. B.* 146:166.

Gale, E., 1962. *J. Appl. Bacteriol.* 25:309.

Gallin, I., and Van Demark, P., 1964. *Biochem. Biophys. Res. Commun.* 17:630.

Ganesan, A., and Lederberg, J., 1965. *Biochem. Biophys. Res. Commun.* 18:824.

Geller, D., 1962. *J. Biol. Chem.* 237:2947.

Gel'man, N. S., 1959. *Usp. Sovrem. Biol.* 47:152.

Gel'man, N. S., 1963. *Enzymes of Electron (Hydrogen) Transfer and Their Connection with the Structure of the Bacterial Cell,* Doctoral Dissertation.

Gel'man, N. S., and Lukoyanova, M. A., 1962. *Mikrobiologiya* 31:556.

Gel'man, N. S., Lukoyanova, M. A., Zhukova, I. G., and Oparin, A. I., 1963. *Biokhimiya* 28:801.

Gel'man, N. S., Zhukova, I. G., Lukoyanova, M. A., and Oparin, A. I., 1959. *Biokhimiya* 24:481.

Gel'man, N. S., Zhukov·:, I. G., and Oparin, A. I., 1959. *Biokhimiya* 24:1074.

Gel'man, N. S., Zhukova, I. G., and Oparin, A. I., 1960*a. Dokl. Akad. Nauk SSSR* 135: 200.

Gel'man, N. S., Zhukova, I. G., and Oparin, A. I., 1960*b. Dokl. Akad. Nauk SSSR* 133: 1209.

Gel'man, N. S., Zhukova, I. G., and Oparin, A. I., 1963. *Biokhimiya* 28:122.

Gel'man, N. S., Zhukova, I. G., and Zaitseva, N. I., 1962. *Dokl. Akad. Nauk SSSR* 145: 206.

Gent, W., Gregsack, N., Gammack, D., and Raper, J., 1964. *Nature* 204:553.

Gershanovich, V. N., Avdeeva, A. V., and Gol'dfarb, D. M., 1963. *Biokhimiya* 28:700.

Gershanovich, V. N., Andreeva, I. V., Burd, G. I., and Zuev, V. A., 1966. *Mikrobiologiya* 25:132.

Gest, H., San Pietro, A., and Vernon, L. (eds.) 1963. *Bacterial Photosynthesis, A Symposium,* Antioch Press, Yellow Springs, Ohio.

Gibbons, R., and Engle, L., 1964. *Science* 146:1307.

Gibbons, R., and Macdonald, J., 1960. *J. Bacteriol.* 80:164.

Gibson, J., 1961. *Biochem. J.* 79:151.

Giesbrecht, P., 1960. *Zentr. Bakteriol.* 179:460.

Giesbrecht, P., 1962. *Zentr. Bakteriol.* 187:452.

Giesbrecht, P., and Drews, G., 1962. *Arch. Mikrobiol.* 43:152.

Gilby, A., and Few, A., 1957. *Second International Congress on Surface Activity,* Vol. 4, London, p. 262.

Gilby, A., and Few, A., 1958. *Nature* 181:55.

Gilby, A., and Few, A., 1960. *Australian J. Biol. Sci.* 13:5.

Gilby, A., Few, A., and McQuillen, K., 1958. *Biochim. Biophys. Acta* 29:21.

Glauert, A., 1962. *Brit. Med. Bull.* 18:245.

Glauert, A., Brieger, E., and Allen, J., 1961. *Exptl. Cell Res.* 22:73.

Glauert, A., and Hopwood, D., 1960. *J. Biochem. Biophys. Cytol.* 7:479.

Glauert, A., Kerridge, D., and Horne, R., 1963. *J. Cell Biol.* 18:327.

Godson, G., and Butler, J., 1962. *Nature* 193:655.

Godson, G., and Butler, J., 1964. *Biochem. J.* 93:573.

Godson, G., Hunter, G., and Butler, J., 1961. *Biochem. J.* 81:59.

Goldfine, H., and Ellis, M., 1964. *J. Bacteriol.* 87:8.

Goldman, D., Wagner, M., Oda, T., and Shug, A., 1963. *Biochim. Biophys. Acta* 73:367.

Gordon, M., and Edwards, M., 1963. *J. Bacteriol.* 86:1101.

Grassowicz, N., and Ariel, M., 1963. *J. Bacteriol.* 85:293.

Gray, C., Wimpenny, J., Hughes, D., and Mossman, M., 1966. *Biochim. Biophys. Acta* 117:22.

Gray, G., and Wilkinson, S., 1965. *J. Appl. Bacteriol.* 28:153.

Green, D., 1962a. Structure and function of subcellular particles. Plenary Lecture. *Proceedings of the Fifth International Biochemical Congress* (Russian edition), USSR Academy of Sciences Press, Moscow. Available in English as: *International Congress of Biochemistry, 5th, Moscow, 1961. Biochemistry: Proceedings,* 9 vols., Sissakian, N. M. (ed.), Pergamon Press, London.

Green, D., 1962b. In: *Structural Components of the Cell* (Russian translation), Foreign Language Press, Moscow, p. 76.

Green, D., 1962c. *J. Comp. Biochem. Physiol.* 4:8.

Green, D., 1964a. In: *Structure and Function of the Cell* (Russian translation), Mir Press, Moscow, p. 216.

Green, D., 1964b. In: *Molecular Biology* (Russian translation), Nauka Press, Moscow, p. 260.

Green, D., Beyer, R., Hansen, M., Smith, A., and Webster, G., 1963. *Federation Proc.* 22:1460.

Green, D., and Fleischer, S., 1962. In: Kasha, M., and Pullman, B. (eds.), *Horizons in Biochemistry,* Academic Press, Inc., New York.

Green, D., and Fleischer, S., 1963. *Biochim. Biophys. Acta* 70:554.

Green, D., and Fleischer, S., 1964. In: Kasha, M., and Pullman, B. (eds.), *Horizons in Biochemistry* (Russian translation), Mir Press, Moscow, p. 293. Originally published in English by Academic Press, Inc., New York, 1962.

Green, D., and Hatefi, Y., 1961. *Science* 133:13.

Green, D., and Hechter, O., 1965. *Proc. Natl. Acad. Sci. U.S.* 53:318.

Green, D., and Lester, R., 1959. *Federation Proc.* 18:987.

Green, D., and Oda, T., 1961. *J. Biochem.* 49:742.

Green, D., and Perdue, J., 1966. *Proc. Natl. Acad. Sci. U.S.* 15:1295.

Green, D., and Wharton, D., 1963. *Biochem. Z.* 338:335.

Greville, G., Munn, E., and Smith, D., 1964. *Proc. Roy. Soc. (London), Ser. B* 161:403.

Griffiths, D., and Beinert, M., 1961. *Biochem. J.* 81:42P.

Griffiths, D., and Wharton, D., 1961. *Biochim. Biophys. Res. Commun.* 4:151, 199.

Grund, S., 1963. *Zentr. Bakteriol.* 189:405.
Gurban, C., and Cristea, E., 1965. *Biochim. Biophys. Acta* 96:195.
Gutnick, D., and Brodie, A., 1965. *J. Biol. Chem.* 240:PC3698.

Hadjipetrou, L., and Stouthamer, A., 1965. *J. Gen. Microbiol.* 38:29.
Hagen, P., Goldgine, H., and Williams, P., 1966. *Science* 151:1543.
Hallberg, P., and Hauge, J., 1965. *Biochim. Biophys. Acta* 95:80.
Harold, F., 1964. *J. Bacteriol.* 88:1416.
Hartman, P., Brodie, A., and Gray, C., 1957. *J. Bacteriol.* 74:319.
Hatefi, Y., 1963. *Advan. Enzymol.* 25:275.
Hatefi, Y., Haavik, A., and Griffiths, D., 1961. *Biochem, Biophys. Res. Commun.* 4:441.
Hatefi, Y., Haavik, A., and Griffiths, D., 1962. *J. Biol. Chem.* 237:1676.
Hauge, J., 1960. *Biochim. Biophys. Acta* 45:250.
Hauge, J., 1961a. *J. Bacteriol.* 82:609.
Hauge, J., 1961b. *Arch. Biochem. Biophys.* 94:308.
Hauge, J., 1964. *J. Biol. Chem.* 239:3630.
Hauge, J., and Hallberg, P., 1964. *Biochim. Biophys. Acta* 81:251.
Hauge, J., and Mürer, E., 1964. *Biochim. Biophys. Acta* 81:244.
Hechter, G., 1965. *Federation Proc.* 24(2) part III:91.
Heinen, W., Kusunose, M., Kusunose, E., Goldman, D., and Wagner, J., 1964. *Arch. Biochem. Biophys.* 104:448.
Hempfling, W., Steinberg, S., and Estabrook, R., 1964. *6th Intern. Congr. Biochem. Abstr. (New York)* X:779.
Hempfling, W., and Vishniac, W., 1965. *Biochem. Z.* 342:272.
Henneman, D., and Umbreit, W., 1964. *J. Bacteriol.* 87:1274.
Heredia, C., and Medina, A., 1960. *Biochem. J.* 77:24.
Hersey, D., and Ajl, S., 1951. *J. Gen. Physiol.* 34:295.
Heydeman, M., and Azoulay, E., 1963. *Biochim. Biophys. Acta* 77:545.
Hickman, D., and Frenkel, A., 1965a. *J. Cell Biol.* 25:279.
Hickman, D., and Frenkel, A., 1965b. *J. Cell Biol.* 25:261.
Hill, P., 1962. *Biochim. Biophys. Acta.* 27:386.
Hines, W., Freeman, B., and Pearson, G., 1964. *J. Bacteriol.* 87:1492.
Hochster, R., and Nozzollio, C., 1959. *Can. J. Biochem. Physiol.* 38:79.
Hockenhull, D., 1948. *Nature* 162:850.
Hoffmann, K., 1962. *Fatty Acid Metabolism in Microorganisms,* Wiley, New York.
Hollocher, T., and Weber, M., 1965. *J. Biol. Chem.* 240:1783.
Holt, S., and Marr, A., 1965a. *J. Bacteriol.* 89:1402.
Holt, S., and Marr, A., 1965b. *J. Bacteriol.* 89:1421.
Hooper, A., and Nason, A., 1965. *J. Biol. Chem.* 240:4044.
Horecker, B., 1962. In: *Evolutionary Biochemistry* (Russian edition), Proceedings of the Fifth International Biochemical Congress, Symposium III, USSR Academy of Sciences Press, Moscow, p. 91. Available in English as: *International Congress of Biochemistry, 5th, Moscow, 1961. Biochemistry: Proceedings,* 9 vols. Sissakian, N. M. (ed.), Pergamon Press, London.
Hori, K., 1961. *J. Biochem.* 50:440.
Hori, K., 1963. *J. Biochem.* 53:354.
Horie, S., and Morrison, M., 1963. *J. Biol. Chem.* 238:1855.
Horie, S., and Morrison, M., 1964. *J. Biol. Chem.* 239:1138.
Horio, T., 1958a. *J. Biochem.* 45:195.
Horio, T., 1958b. *J. Biochem.* 45:267.
Horio, T., Higashi, T., Matsubara, H., Kusai, K., Nakai, M., and Okunuki, K., 1958a. *Biochim. Biophys. Acta* 29:297.

Horio, T., Higashi, T., Nakai, M., Kusai, K., and Okunuki, K., 1958b. *Nature* 182:1307.
Horio, T., Higashi, T., Sasagawa, T., Kusai, K., Nakai, M., and Okunuki, K., 1960. *Biochem. J.* 77:194.
Horio, T., Higashi, T., Yamanaka, M., Matsubara, H., and Okunuki, K., 1961a. *J. Biol. Chem.* 236:944.
Horio, T., and Kamen, M., 1961. *Biochim. Biophys. Acta* 48:266.
Horio, T., Sekuzu, I., Higashi, T., and Okunuki, K., 1961b. In: Falk, J., Lemberg, R., and Morton, R. (eds.), *Hematin Enzymes,* Vol. I, Pergamon Press, Oxford, p. 302.
Horio, T., and Taylor, C., 1965. *J. Biol. Chem.* 240:1772.
Houtsmuller, U., and Van Deenen, L., 1963. *Biochem. J.* 88:43P.
Houtsmuller, U., and Van Deenen, L., 1964. *Biochim. Biophys. Acta* 84:96.
Houtsmuller, U., and Van Deenen, L., 1965. *Biochim. Biophys. Acta* 106:564.
Hovenkamp, H., 1959a. *Nature* 184:471.
Hovenkamp, H., 1959b. *Biochim. Biophys. Acta* 34:485.
Huang, C., Wheeldon, L., and Thompson, T., 1964. *J. Mol. Biol.* 8:148.
Hughes, D., 1961. *J. Biochem. Microbiol. Technol. Eng.* 3:405.
Hughes, D., 1962. *J. Gen. Microbiol.* 29:39.
Hughes, D., and Cunningham, V., 1963. *Biochem. Soc. Symp.* 23:8.
Hughes, D., Hunt, A., Rodgers, A., and Lowenstein, J., 1959. In: Hoffmann-Ostenhof, O. (ed.), *International Congress of Biochemistry, 4th, Vienna, 1958. Biochemistry: Proceedings,* 15 vols., Macmillan, New York.
Hunt, A., Rodgers, A., and Hughes, D., 1959. *Biochim. Biophys. Acta* 34:354.
Hunter, G., and Goodsall, R., 1961. *Biochem. J.* 78:564.
Huston, C., and Albro, P., 1964. *J. Bacteriol.* 88:425.
Hyndman, L., Burris, R., and Wilson, P., 1953. *J. Bacteriol.* 65:522.

Ibbott, F., and Abrams, A., 1964a. *Federation Proc.* 23:222.
Ibbott, F., and Abrams, A., 1964b. *Biochemistry* 3:2008.
Ierusalimskii, N. D., 1963. *Fundamentals of Microbial Physiology,* USSR Academy of Sciences Press, Moscow.
Iida, K., and Taniguchi, S., 1959. *J. Biochem.* 46:1041.
Ikawa, M., 1963. *J. Bacteriol.* 85:773.
Imaeda, T., and Convit, J., 1962. *J. Bacteriol.* 83:43.
Imaeda, T., and Ogura, M., 1963. *J. Bacteriol.* 85:150.
Imshenetskii, A. A., 1962. In: *Evolutionary Biochemistry,* Proceedings of the Fifth International Biochemical Congress, Symposium III (Russian edition), USSR Academy of Sciences Press, Moscow, p. 141. Available in English as: *International Congress of Biochemistry, 5th, Moscow, 1961. Biochemistry: Proceedings,* 9 vols., Sissakian, N. M. (ed.), Pergamon Press, London.
Ishikawa, S., and Lehninger, A., 1962. *J. Biol. Chem.* 237:2401.
Ishimoto, M., and Egami, F., 1957. In: *The Origin of Life on Earth,* Oparin, A. I., *et al.* (eds.) (Russian edition), USSR Academy of Sciences Press, Moscow, p. 326. Published in English as *International Symposium on the Origin of Life on the Earth, Moscow. Origin of Life on Earth: Proceedings,* Pergamon Press, London.
Ishimoto, M., and Fujimoto, D., 1959. *Proc. Japan Acad.* 35:243.
Ishimoto, M., Koyama, J., Ohmura, T., and Nagai, Y., 1954. *J. Biochem.* 41:537.
Ishimoto, M., Koyama, J., Yagi, T., and Shiraki, M., 1957. *J. Biochem.* 44:413.
Ishimoto, M., and Yagi, T., 1961. *J. Biochem.* 49:103.
Itagaki, E., 1964. *J. Biochem.* 55:432.
Itagaki, E., Fujita, T., and Sato, R., 1961. *Biochim. Biophys. Acta* 51:380.
Itagaki, E., and Taniguchi, S., 1959. *J. Biochem.* 45:1419.
Ito, S., and Vinson, J., 1965. *J. Bacteriol.* 89:481.

Iwasaki, Y., 1960. *Plant Cell Physiol.* 1:195, 207.

Jackson, F., and Lawton, V., 1958. *Nature* 181:1539.
Jackson, F., and Lawton, V., 1959. *Biochim. Biophys. Acta* 35:76.
Jacob, R., Ryter, A., and Cuzin, F., 1966. *Proc. Roy. Soc. (London), Ser. B* 164:267.
Jacobs, N., and Conti, S., 1965. *J. Bacteriol.* 89:675.
Jacobs, N., and Wolin, M., 1963. *Biochim. Biophys. Acta* 69:18.
Jacobsen, B., and Dam, H., 1960. *Biochim. Biophys. Acta* 40:211.
Jones, C., and Benson, A., 1965. *J. Bacteriol.* 89:260.
Jones, C., and King, H., 1964. *Biochem. J.* 91:10P.
Jones, C., and Redfearn, E., 1966. *Biochim. Biophys. Acta* 113:467.
Jose, A., and Wilson, A., 1959. *Proc. Natl. Acad. Sci. U.S.* 45:692.
Judah, I., and Ahmed, K., 1964. *Biol. Rev.* 39:160.
Juhasz, S., 1961. *Can. J. Microbiol.* 7:832.
Jurtshuk, P., Sekuzu, I., and Green, D., 1961. *Biochem. Biophys. Res. Commun.* 6:76.
Jurtshuk, P., Sekuzu, I., and Green, D., 1963. *J. Biol. Chem.* 238:3595.

Kagawa, Y., and Racker, E., 1965. *Federation Proc.* 24(2) part 1:363.
Kamen, M., 1962. In: *Mechanism of Photosynthesis* (Russian edition), Proceedings of the Fifth International Biochemical Congress, Symposium V, USSR Academy of Sciences Press, Moscow, p. 253. Available in English as: *International Congress of Biochemistry, 5th, Moscow, 1961. Biochemistry: Proceedings*, 9 vols., Sissakian, N. M. (ed.), Pergamon Press, London.
Kamen, M., 1963. In: Gest, H., San Pietro, A., and Vernon, L. (eds.), *Bacterial Photosynthesis*, Antioch Press, Yellow Springs, Ohio, p. 61.
Kamen, M., Bartsch, R., Horio, T., and de Klerk, H., 1963. In: Colowick, S., and Kaplan, N. (eds.), *Methods in Enzymology*, Vol. 6, Academic Press, Inc., New York, p. 391.
Kamen, M., and Vernon, L., 1954a. *J. Bacteriol.* 67:617.
Kamen, M., and Vernon, L., 1954b. *J. Biol. Chem.* 211:663.
Kamen, M., and Vernon, L., 1955. *Biochim. Biophys. Acta* 17:10.
Kaneshiro, R., and Marr, A., 1961. *J. Biol. Chem.* 236:2615.
Kaneshiro, R., and Marr, A., 1962. *J. Lipid Res.* 3:184.
Karpenko, M. K., Kvasnikov, E. I., and Burakova, A. A., 1964. *Mikrobiol. Zh.* 26:6.
Kashket, E., and Brodie, A., 1960. *Biochim. Biophys. Acta* 40:550.
Kashket, E., and Brodie, A., 1963a. *Biochim. Biophys. Acta* 78:52.
Kashket, E., and Brodie, A., 1963b. *J. Biol. Chem.* 238:2564.
Kates, M., 1964. *Advan. Lipid Res.* 2:17.
Kauzmann, W., 1959. *Advan. Protein Chem.* 14:1.
Kavanau, J., 1963. *Nature* 198:525.
Kavanau, J., 1965. *Structure and Function in Biological Membranes*, 2 vols., Holden-Day, Inc., San Francisco.
Kawata, T., 1963. *J. Gen. Appl. Microbiol.* 9:1.
Kawata, T., and Inoue, T., 1964. *Japan J. Microbiol.* 8:49.
Kawata, T., Inoue, T., and Takagi, A., 1963. *Japan J. Microbiol.* 7:23.
Keilin, D., and Harpley, C., 1941. *Biochem. J.* 35:688.
Keilin, D., and Hartree, E., 1938. *Nature* 141:870.
Keilin, D., and Hartree, E., 1939. *Proc. Roy. Soc. (London), Ser. B* 127:167.
Kellenberger, E., and Ryter, A., 1958. *J. Biophys. Biochem. Cytol.* 4:323.
Kellenberger, E., and Ryter, A., 1964. In: Siegel, B. M. (ed.), *Modern Developments in Electron Microscopy*, Academic Press, Inc., New York, p. 335.
Kennedy, E., 1964. *6th Internat. Congr. Biochem. Abstr. (New York)* VII:555.

Kerkut, K., 1960. *Implications of Evolution,* Vol. 4, Pergamon Press, London.
Kersters, G., Wood, W., and De Ley, J., 1965. *J. Biol. Chem.* 240:965.
Kidwai, A., and Murti, K., 1965. *Ind. J. Biochem.* 2:217.
Kiesow, L., 1964. *Proc. Natl. Acad. Sci. U.S.* 52:980.
Kikuchi, G., Saito, Y., and Motokawa, Y., 1965. *Biochim. Biophys. Acta* 94:1.
Kimura, T., and Tobari, J., 1963. *Biochim. Biophys. Acta* 73:399.
King, H., Bishop, D., and Pandya, K., 1961. In: *Abstracts of Sectional Papers of the Fifth International Biochemical Congress* (Russian edition), Vol. II, USSR Academy of Sciences Press, Moscow, p. 393. Available in English as: *International Congress of Biochemistry, 5th, Moscow, 1961. Biochemistry: Proceedings,* 9 vols., Sissakian, N. M. (ed.), Pergamon Press, London.
King, T., and Cheldelin, V., 1957. *J. Biol. Chem.* 224:579.
King, T., and Cheldelin, V., 1958. *Biochem. J.* 68:31P.
King, T., and Kuboyama, M., 1964. *Biochem. Biophys. Res. Commun.* 17:231.
King, T., Nickel, K., and Jensen, D., 1964. *J. Biol. Chem.* 239:1989.
Kleczowska, H., Wiaterowa, A., and Bagdasarian, G., 1965. *Acta Microbiol. Polon.* 14: 117.
Kline, E., and Mahler, H., 1965. *Ann. N. Y. Acad. Sci.* 119:905.
Klingenberg, M., and Schollmeyer, P., 1962. In: *Intercellular Respiration, Phosphorylating and Nonphosphorylating Oxidation Reactions* (Russian edition), Proceedings of the Fifth International Biochemical Congress, Symposium V, USSR Academy of Sciences Press, Moscow, p. 55. Available in English as *International Congress of Biochemistry, 5th, Moscow, 1961. Biochemistry: Proceedings,* 9 vols., Sissakian, N. M. (ed.), Pergamon Press, London.
Klots, I., 1964. In: *Horizons in Biochemistry,* Kasha, M., and Pullman, B. (eds.), (Russian translation), Mir Press, Moscow, p. 399. Originally published in English by Academic Press, Inc., New York, 1962.
Kluyver, A., and Van Niel, C., 1959. *The Microbe's Contribution to Biology* (Russian translation), Foreign Literature Press, Moscow. Originally published in English by Harvard University Press, Cambridge, Mass.
Klüngsoÿr, L., King, T., and Cheldelin, V., 1957. *J. Biol. Chem.* 227:135.
Kodicek, E., 1963. In: Gibbons, N. (ed.), *Recent Progress in Microbiology,* University of Toronto Press, Toronto, p. 23.
Koike, M., and Takeya, K., 1961. *J. Biophys. Biochem. Cytol.* 9:597.
Koivusalo, M., Currie, R., and Slater, E., 1966. *Biochim. Z.* 344:221.
Kondrat'eva, E. N., 1963. *Photosynthetic Bacteria,* USSR Academy of Sciences Press, Moscow.
Kondrat'eva, E. N., 1964. *Usp. Mikrobiol.* 1:5.
Kono, T., and Colowick, S., 1961. *Arch. Biochem. Biophys.* 93:520.
Kornberg, H., and Phizackerley, P., 1961. *Biochem. J.* 79:10P.
Kotel'nikova, A. V., 1962. *Usp. Biol. Khim.* 4:173.
Kotel'nikova, A. V., 1964. *Usp. Biol. Khim.* 6:156.
Kotel'nikova, A. V., and Ivanova, E. V., 1964. *Dokl. Akad. Nauk SSSR* 157:710.
Kran, K., 1962. *Arch. Mikrobiol.* 44:1.
Krasnovskii, A. A., 1957. In: The Origin of Life on Earth, Oparin, A. I., *et al.* (eds.) (Russian edition), SSSR Academy of Sciences Press, Moscow, p. 355. Published in English as *International Symposium on the Origin of Life on the Earth, Moscow. Origin of Life on Earth: Proceedings,* Pergamon Press, London.
Krogstad, D., and Howland, J., 1966. *Biochim. Biophys. Acta* 118:189.
Kroon, A., 1964. *Biochim. Biophys. Acta* 91:145.
Kurup, C., and Vaidyanathan, C., 1964. *Ind. J. Biochem.* 1:111.
Kurup, C., Vaidyanathan, C., and Ramasarma, T., 1966. *Arch. Biochem. Biophys.* 113: 548.
Kushnarev, V. M., 1962. *Byull. Eksperim. Biol. i Med.* 3:65.

Kushnarev, V. M., 1966. *Dokl. Akad. Nauk SSSR* (in press).
Kushnarev, V. M., and Pereverzev, N. A., 1964. *Mikrobiologiya* 33:610.
Kusunose, E., and Goldman, D., 1963. *Biochim. Biophys. Acta* 73:391.
Kusunose, M., and Kusunose, E., 1959. *Ann. Rept. Japan Soc. Tuberc.* 4:17.
Kuznetsov, V. P., 1965. *Antibiotiki* 1:58.

Lara, T., 1959. *Biochim. Biophys. Acta* 33:565.
Lark, K., 1966. *Bacteriol. Rev.* 30:3.
Lascelles, J., 1965. In: *Structure and Function in Microorganisms,* Vol. 15, Soc. Gen. Microbiol., p. 32.
Lascelles, J., and Szilágyi, J., 1965. *J. Gen. Microbiol.* 38:55.
Layne, E., and Nason, A., 1958. *J. Biol. Chem.* 231:889.
Lederer, E., 1964. *Biochem. J.,* 93:449.
Leene, W., and Van Iterson, W., 1965. *J. Cell Biol.* 27:241.
Lees, H., 1958. *Biochemistry of Autotrophic Bacteria* (Russian translation), Foreign Language Press, Moscow.
Lees, H., 1960. *Ann. Rev. Microbiol.* 14:83.
Lees, H., and Simpson, J., 1957. *Biochem. J.* 65:297.
LeGall, J., Mazza, G., and Dragoni, N., 1965. *Biochim. Biophys. Acta* 99:385.
Lehninger, A., 1959. *Rev. Modern Phys.* 31:137.
Lehninger, A., 1964. *The Mitochondrion,* Benjamin Press, New York.
Lemberg, R., Clezy, P., and Barret, J., 1961. In: Falk, J., Lemberg, R., and Morton, R. (eds.), *Hematin Enzymes,* Vol. 1, Pergamon Press, Oxford, p. 344.
Lemberg, R., Newton, N., and O'Hagan, J., 1962. *Proc. Roy. Soc. (London), Ser. B* 159:405.
Lenaz, G., and Beyer, R., 1965a. *Federation Proc.* 24:363.
Lenaz, G., and Beyer, R., 1965b. *J. Biol. Chem.* 240:3653.
Lennarz, W., 1964. *J. Biol. Chem.* 239:PC3111.
Leonova, N. G., Mokhova, E. N., and Pomushchnikova, N. A., 1966. *Mikrobiologya* 35:944.
Lester, R., and Crane, F., 1959. *J. Biol. Chem.* 234:2169.
Lester, R., White, D., and Smith, S., 1964. *Biochemistry* 3:949.
Lev, M., 1959. *J. Gen. Microbiol.* 20:697.
Lightbown, J., and Jackson, F., 1956. *Biochem. J.* 63:130.
Linnane, A., and Wrigley, C., 1963. *Biochim. Biophys. Acta* 77:408.
Lipman, F., 1966. In: *The Origins of Prebiological Systems,* Fox, S. W. (ed.) (Russian translation), Mir Press, Moscow. Originally published in English by Academic Press, Inc., New York.
Lisenkova, L. L., 1965. Personal communication.
Lisenkova, L. L., and Lozinov, A. B., 1966. *Prikl. Biokhim. i Mikrobiol.* 2:175.
Lisenkova, L. L., and Mokhova, E. N., 1964. *Mikrobiologiya* 33:918.
Listgarten, M., Loesche, W., and Socransky, S., 1963. *J. Bacteriol.* 85:933.
London, J., 1963. *Science* 140:409.
Löw, H., and Afzelius, B., 1964. *Exptl. Cell Res.* 35:431.
Lozinov, A. B., and Ermachenko, V. A., 1962. *Mikrobiologiya* 31:927.
Lucy, J., 1964. *J. Theoret. Biol.* 7:360.
Lucy, J., and Glauert, A., 1964. *J. Mol. Biol.* 8:727.
Ludýik, J., 1964. In: *Electron Microscopy,* Czechoslovakian Academy of Sciences, Prague, p. 533.
Lukoyanova, M. A., 1964. *Nature of the Organization of Respiratory Enzymes of Membranes in M. lysodeikticus,* Candidate Dissertation.
Lukoyanova, M. A., and Biryuzova, V. I., 1965. *Biokhimiya* 30:599.
Lukoyanova, M. A., Gel'man, N. S., and Biryuzova, V. I., 1961. *Biokhimiya* 26:916.
Lukoyanova, M. A., Gel'man, N. S., and Oparin, A. I., 1963. *Biokhimiya* 28:334.

Luria, S., 1960. In: Gunsalus, I., and Stanier, R. (eds.), *Bacteria: A Treatise on Structure and Function,* Vol. 1, Academic Press, Inc., New York, p. 1.
Lusena, C., and Dass, C., 1963. *Can. J. Biochem. Physiol.* 41:2205.
Luzzatti, V., and Husson, F., 1962. *J. Cell Biol.* 12:207.
Lyalikova, N. N., 1958. *Mikrobiologiya* 27:556.
Lynn, W., and Brown, R., 1965. *Biochim. Biophys. Acta* 105:15.

Macfarlane, M., 1961*a. Biochem. J.* 79:4P.
Macfarlane, M., 1961*b. Biochem. J.* 80:45P.
Macfarlane, M., 1962*a. Biochem. J.* 82:40P.
Macfarlane, M., 1962*b. Nature* 196:136.
Macfarlane, M., 1964*a. Advan. Lipid Res.* 2:91.
Macfarlane, M., 1964*b. 6th Internat. Congr. Biochem. Abstr. (New York)* VII:551.
MacLennan, D., Tzagoloff, A., and Rieske, J., 1965. *Arch. Biochem. Biophys.* 109:383.
MacLeod, R., 1965. *Bacteriol. Rev.* 29:9.
Madsen, N., 1960. *Can. J. Biochem. Physiol.* 38:481.
Maeno, H., 1965. *J. Biochem.* 57:657.
Malatyan, M. N., 1962. *Dokl. Akad. Nauk SSSR* 143:955.
Malatyan, M. N., 1963. *Mikrobiologiya* 32:806.
Malatyan, M. N., and Biryuzova, V. I., 1965. *Dokl. Akad. Nauk SSSR* 160:1182.
Malavolta, E., Delwiche, C., and Burge, M., 1960. *Biochem. Biophys. Res. Commun.* 2: 445.
Mal'tseva, N. N., 1963. *Physiological Characteristics of B. megatherium De Bary,* Candidate Dissertation, Kiev.
Marinetti, G.. 1964. *Biochim. Biophys. Acta* 84:55.
Markowitz, A., and Dorfman, A., 1962. *J. Biol. Chem.* 237:273.
Marquis, R., 1965. *J. Bacteriol.* 89:1453.
Marquis, R., and Gerhardt, P., 1964. *J. Biol. Chem.* 239:3361.
Marr, A., 1960. *Ann. Rev. Microbiol.* 14:253.
Martin, H., 1963. *J. Theoret. Biol.* 5:1.
Martius, C., and Nitz-Litzow, D., 1953. *Biochim. Biophys. Acta* 12:152.
Martius, C., and Nitz-Litzow, D., 1954. *Biochim. Biophys. Acta* 13:152.
Mathews, M., and Sistrom, W., 1959. *J. Bacteriol.* 78:778.
Matsubara, H., Orii, Y., and Okunuki, K., 1965. *Biochim. Biophys. Acta* 97:61.
McEwen, B., Allfrey, V., and Mirsky, A., 1963. *J. Biol. Chem.* 238:758.
Meisel', M. N., Biryuzova, V. I., Volkova, T. M., Malatyan, M. N., and Medvedeva, G. A., 1964. In: *Electron and Fluorescence Microscopy,* Nauka Press, Moscow, p. 3.
Meisel', M. N., and Mirolyubova, L. V., 1959. *Izv. Akad. Nauk SSSR, Ser. Biol.* 6:865.
Merrick, I., and Doudoroff, M., 1961. *Nature* 189:4768.
Mikhlin, D. M., 1960. *Biochemistry of Cell Respiration,* USSR Academy of Sciences Press, Moscow.
Minakami, S., Schindler, R., and Estabrook, R., 1964. *J. Biol. Chem.* 239:2042.
Mitchell, P., 1959*a. Biochem. Soc. Symp.* 16:73.
Mitchell, P., 1959*b. Ann. Rev. Microbiol.* 13:407.
Mitchell, P., 1961*a. Biochem. J.* 81:24P.
Mitchell, P., 1961*b. Nature* 191:48.
Mitchell, P., 1962. *J. Gen. Microbiol.* 29:25.
Mitchell, P., 1963. In: McLaren, A. D. (ed.), *Cell Interface Reactions,* Scholar's Library, New York.
Mitchell, P., and Moyle, J., 1956. *J. Gen. Microbiol.* 15:512.
Mitchell, P., and Moyle, J., 1959. *Biochem. J.* 72:21P.

Mitchell, P., and Moyle, J., 1965. *Nature* 208:147.
Mitsuhashi, S., Kojima, Y., and Yagi, K., 1956. *J. Biochem.* 43:337.
Mitsui, H., 1961. *J. Gen. Appl. Microbiol.* 7:262.
Mizuno, S., Yoshida, E., Takahashi, H., and Maruo, B., 1961. *Biochim. Biophys. Acta* 49:369.
Mizushima, S., and Arima, K., 1960. *J. Biochem.* 47:837.
Mokhova, E. N., 1965. *Biofizika* 10:571.
Moore, R., and McAlear, J., 1961. *Exptl. Cell Res.* 24:588.
Morita, S., 1960. *J. Biochem.* 48:870.
Morita, S., and Conti, S., 1963. In: Gest, H., San Pietro, A., and Vernon, L. (eds.), *Bacterial Photosynthesis,* Antioch Press, Yellow Springs, Ohio, p. 327.
Morowitz, H., and Tourtellotte, M., 1964. In: *The Structure and Function of the Living Cell* (Russian translation), Foreign Literature Press, Moscow, p. 104.
Morowitz, H., Tourtellotte, M., Guild, W., Castro, E., and Woese, C., 1962. *J. Mol. Biol.* 4:93.
Morrison, M., and Stotz, E., 1961. In: Falk, J., Lemberg, R., and Morton, R. (eds.), *Hematin Enzymes,* Pergamon Press, Oxford, p. 335.
Moss, F., 1956. *Australian J. Exptl. Biol. Med. Sci.* 34:395.
Mudd, S., Kawata, T., Payne, J., Sall, T., and Takagi, A., 1961. *Nature* 189:79.
Murray, R., 1960. In: Gunsalus, I., and Stanier, R. (eds.), *Bacteria: A Treatise on Structure and Function,* Vol. I, Academic Press, New York, p. 35.
Murray, R., and Birch-Andersen, A., 1963. *Can. J. Microbiol.* 9:393.
Murray, R., and Watson, S., 1965. *J. Bacteriol.* 89:1594.
Murti, K., 1960. *Biochim. Biophys. Acta* 45:243.

Nadakavukaren, M., 1964. *J. Cell Biol.* 23:193.
Naik, M., and Nicholas, D., 1966. *Biochim. Biophys. Acta* 113:490.
Nakada, D., Nozu, K., and Kondo, M., 1958. *J. Biochem.* 45:751.
Nakayama, T., 1961. *J. Biochem.* 49:240.
Nakayama, T., and De Ley, J., 1965. *Antonie van Leeuwenhoek, J. Microbiol. Serol.* 31: 205.
Nason, A., 1962. *Bacteriol. Rev.* 26:16.
Nason, A., and Aleem, M., 1961. In: *Abstracts of Sectional Papers of the Fifth International Biochemical Congress* (Russian translation), Vol. II, USSR Academy of Sciences Press, Moscow, p. 402. Available in English as: *International Congress of Biochemistry, 5th, Moscow, 1961. Biochemistry: Proceedings,* 9 vols. Sissakian, N. M. (ed.), Pergamon Press, London.
Nass, M., Nass, S., and Afzelius, B., 1965. *Exptl. Cell Res.* 37:516.
Nermut, M., 1960. *Zentr. Bakteriol.* 178:348.
Nermut, M., 1963. *Folia Microbiol. (Prague)* 8:370.
Nermut, M., 1964. *Z. Allgem. Mikrobiol.* 4:92.
Nermut, M., 1965. *Folia Microbiol. (Prague)* 10:104.
Nermut, M., and Rýc, M., 1964. *Folia Microbiol. (Prague)* 9:16.
Newton, B., 1960. In: *The Strategy of Chemotherapy,* Cowan, S. T., and Rowatt, E. (eds.) (Russian translation), Foreign Literature Press, Moscow, p. 15. Originally published in English by Cambridge University Press, New York.
Newton, J., and Kamen, M., 1963. In: Gusalus, I., and Stainer, R. (eds.), *Bacteria: A Treatise on Structure and Function,* Vol. II (Russian translation), Foreign Literature Press, Moscow. Originally published in English by Academic Press, Inc., New York, in 5 vols., 1960–1964.
Nicholas, D., 1964. *6th Intern. Congr. Biochem. Abstr. (New York)* X:784.
Nicholas, D., and Rao, P., 1964. *Biochim. Biophys. Acta* 82:394.

Niederpruem, D., and Doudoroff, M., 1965. *J. Bacteriol.* 89:697.

Nielsen, L., and Abrams, A., 1964. *Biochem. Biophys. Res. Commun.* 17:680.

Nikitina, E. S., and Biryuzova, V. I., 1965. *Mikrobiologiya* 34:1008.

Niklowitz, W., 1958. *Zentr. Bakteriol.* 172:12.

Ning Ling, G., 1965. *Federation Proc.* 24(2) part III: 103.

Nishizuka, Y., and Hayashi, O., 1962. *J. Biol. Chem.* 237:2721.

Nishizuka, Y., Kuno, S., and Hayashi, O., 1960. *J. Biol. Chem.* 235:897.

Nisman, B., Zubrzycki, Z., and Pelmont, I., 1963. *Life Sciences* 10:768.

North, R., 1963. *J. Ultrastruct. Res.* 9:187.

Norton, J., Bulmer, G., and Sokatch, J., 1963. *Biochim. Biophys. Acta* 78:136.

Nossal, P., Keech, D., and Morton, D., 1956. *Biochim. Biophys. Acta* 22:412.

Novikoff, A., 1962. In: *Structural Components of the Cell* (Russian translation), Foreign Literature Press, Moscow, p. 15.

Ogura, M., 1963. *J. Ultrastruct. Res.* 8:251.

Ohnishi, T., 1963. *J. Biochem.* 53:71.

Ohnishi, T., and Mori, T., 1960. *J. Biochem.* 48:406.

Ohnishi, T., and Mori, T., 1962. *Nature* 193:488.

Ohye, D., and Murrell, W., 1962. *J. Cell Biol.* 14:111.

Okui, S., Suzuki, Y., and Momose, K., 1963. *J. Biochem.* 54:471.

O'Leary, W., 1962. *Bacteriol. Rev.* 26:421.

Oparin, A. I., 1924. *The Origin of Life,* Moskovskii Rabochii Press, Moscow. English translation: Dover Publications, Inc., 2nd edition, paper, 1953.

Oparin, A. I., 1957. *The Origin of Life on the Earth,* USSR Academy of Sciences Press, Moscow. Published in English by Academic Press, Inc., New York, 1957 (3rd edition).

Oparin, A. I., 1960. *Life: Its Nature, Origin and Development,* USSR Academy of Sciences Press, Moscow. Published in English by Academic Press, Inc., New York.

Oparin, A. I., 1962. In: *Evolutionary Biochemistry* (Russian edition), Proceedings of the Fifth International Biochemical Congress, Symposium III, USSR Academy of Sciences Press, Moscow, p. 74. Available in English as: *International Congress of Biochemistry, 5th, Moscow, 1961. Biochemistry: Proceedings,* 9 vols., Sissakian, N. M. (ed.), Pergamon Press, London.

Oparin, A. I., 1966. In: *The Origins of Prebiological Systems,* Fox, S. W. (ed.) (Russian translation), Mir Press, Moscow (in press). Originally published in English by Academic Press, Inc., New York.

Oparin, A. I., Gel'man, N. S., and Kharat'yan, E. F., 1964. *Dokl. Akad. Nauk SSS* 157: 211.

Oparin, A. I., Gel'man, N. S., and Zhukova, I. G., 1965. *Dokl. Akad. Nauk SSSR* 161: 237.

Oparin, A. I., Gel'man, N. S., Zhukova, I. G., Shvets, V. I., Chergadze, Yu. N., and Tsfasman, I. M., 1963. *Dokl. Akad. Nauk SSSR* 152:228.

Oparin, A. I., Lukoyanova, M. A., Shvets, V. I., Gel'man, N. S., and Torkhovskaya, T. I., 1965. *Zh. Evolyutsion Biokhim. Fiziol.* 1:7.

Oparin, A. I., and Serebrovskaya, K. B., 1963. *Dokl. Akad. Nauk SSSR* 148:943.

Op den Kamp, J., Houtsmuller, U., and Van Deenen, L., 1965. *Biochim. Biophys. Acta* 106:438.

Osnitskaya, L., Threlfall, D., and Goodwin, T., 1964. *Nature* 204:80.

Ostrovskii, D. N., 1964a. *Proceedings of a Scientic Conference on the Determination of Oxygen Tension in Living Tissues by the Polarographic Method, Experimentally and in the Clinic,* Gorky.

Ostrovskii, D. N., 1964*b*. *An Investigation of Oxidative Phosphorylation in Membranes of Micrococcus lysodeikticus,* Candidate Dissertation, Moscow.
Ostrovskii, D. N., and Gel'man, N. S., 1962. *Biokhimiya* 27:532.
Ostrovskii, D. N., and Gel'man, N. S., 1963. *Doklady Akad. Nauk SSSR* 148:945.
Ostrovskii, D. N., and Gel'man, N. S., 1965. *Biokhimiya* 30:772.
Ostrovskii, D. N., Kharat'yan, E. F., and Gel'man, N. S., 1964. *Biokhimiya* 29:154.
Ota, A., 1965. *J. Biochem. (Japan)* 58:137.
Ota, A., Yamanaka, T., and Okunuki, K., 1964. *J. Biochem.* 55:131.
Overman, J., and Pine, L., 1963. *J. Bacteriol.* 86:655.
Oxenburgh, M., and Snoswell, A., 1965. *J. Bacteriol.* 89:913.

Packer, L., 1958*a. Bacteriol. Proc.* p. 106.
Packer, L., 1958*b. Arch. Biochem. Biophys.* 78:54.
Packer, L., and Vishniac, W., 1955. *J. Bacteriol.* 70:216.
Page, A., Gale, P., Wallick, H., Walton, R., McDaniel, L., Woodruff, H., and Folkers, K., 1960. *Arch. Biochem. Biophys.* 89:318.
Palade, G., 1964. *Proc. Natl. Acad. Sci. U.S.* 52:613.
Pandit-Hovenkamp, H., 1964. *6th Intern. Congr. Biochem. Abstr. (New York)* X:785.
Pandit-Hovenkamp, H., 1965. *Biochim. Biophys. Acta* 99:552.
Pandya, K., Bishop, D., and King, H., 1961. *Biochem. J.* 78:35P.
Pandya, K., and King, H., 1962. *Biochem. J.* 85:15P.
Pandya, K., and King, H., 1966. *Arch. Biochem. Biophys.* 114:154.
Pangborn, J., Marr, A., and Robrish, S., 1962. *J. Bacteriol.* 84:669.
Pappenheimer, A., 1955. In: Colowick, S., and Kaplan, N. (eds.), *Methods in Enzymology,* Vol. 2, Academic Press, Inc., New York, p. 744.
Pappenheimer, A., and Hendee, E., 1949. *J. Biol. Chem.* 180:597.
Pappenheimer, A., Howland, J., and Miller, P., 1962. *Biochim. Biophys. Acta* 64:229.
Parsons, D., 1963*a. Science* 140:985.
Parsons, D., 1963*b. J. Cell Biol.* 16:620.
Pavlova, I., 1964. In: *Electron Microscopy.* Czechoslovakian Academy of Sciences, Prague, p. 531.
Peck, H., 1959. *Proc. Natl. Acad. Sci. U.S.* 45:701.
Peck, H., 1962. *Bacteriol. Rev.* 26:67.
Penniall, R., Hall, A., Griffin, S., and Irvin, I., 1962. *Arch. Biochem. Biophys.* 98:128.
Penniall, R., Saunders, I., and Liu, Shin-min, 1964. *Biochemistry* 3:1454, 1459.
Pepper, R., and Costilow, R., 1965. *J. Bacteriol.* 89:271.
Person, P., and Zipper, H., 1965. *Biochem. Biophys. Res. Commun.* 18:396.
Pesch, L., and Peterson, J., 1965. *Biochim. Biophys. Acta* 96:390.
Pethica, B., 1958. *J. Gen. Microbiol.* 18:473.
Petruschka, E., Quastel, J., and Scholefield, P., 1959. *Can. J. Biochem. Physiol.* 37:975, 989.
Pfister, R., and Lundgren, D., 1964. *J. Bacteriol.* 88:1119.
Pinchot, G., 1953. *J. Biol. Chem.* 205:65.
Pinchot, G., 1955. *J. Am. Chem. Soc.* 77:5763.
Pinchot, G., 1957*a. J. Biol. Chem.* 229:1.
Pinchot, G., 1957*b. J. Biol. Chem.* 229:11.
Pinchot, G., 1957*c. J. Biol. Chem.* 229:25.
Pinchot, G., 1957*d. Biochim. Biophys. Acta* 23:660.
Pinchot, G., 1959. *Biochem. Biophys. Res. Commun.* 1:17.
Pinchot, G., 1960. *Proc. Natl. Acad. Sci. U.S.* 46:929.
Pinchot, G., 1963. *Federation Proc.* 22:1076.
Pinchot, G., 1965. *Perspectives Biol. Med.* 8:180.

Pinchot, G., and Hormanski, M., 1962a. *Proc. Natl. Acad. Sci. U.S.* 48:1970.
Pinchot, G., and Hormanski, M., 1962b. *Federation Proc.* 21:54.
Pinchot, G., and Racker, E., 1951. *Federation Proc.* 10:233.
Pirie, N., 1964. *Proc. Roy. Soc. (London), Ser. B* 160:149.
Polglase, W., Pun, W., and Withaar, J., 1966. *Biochim. Biophys. Acta* 118:426.
Postgate, J., 1952. *Research* 5:189.
Postgate, J., 1956. *J. Gen. Microbiol.* 14:545.
Postgate, J., 1959. *Ann. Rev. Microbiol.* 13:505.
Postgate, J., 1961. In: Falk, J., Lemberg, R., and Morton, R. (eds.), *Hematin Enzymes*, Vol. 2, Pergamon Press, Oxford, p. 407.
Postgate, J., 1965. *Bacteriol. Rev.* 29:425.
Prakash, O., Rao, R., and Sadana, J., 1966. *Biochim. Biophys. Acta* 118:426.
Pullman, B., and Pullman, A., 1965. *Quantum Biochemistry* (Russian translation), Mir Press, Moscow, p. 413. Originally published in English by (Interscience) Wiley.
Pumphrey, A., and Redfearn, E., 1960. *Biochem. J.* 76:61.

Racker, E., 1962. *Proc. Natl. Acad. Sci. U.S.* 48:1659.
Racker, E., 1963. *Biochem. Biophys. Res. Commun.* 10:435.
Racker, E., 1964. *Biochem. Biophys. Res. Commun.* 14:75.
Racker, E., 1965. *Mechanisms in Bioenergetics,* Academic Press, Inc., New York.
Racker, E., and Conover, T., 1963. *Federation Proc.* 22:1088.
Racker, E., and Monroy, G., 1964. *6th Intern. Congr. Biochem. Abstr. (New York)* X:760.
Racker, E., Parsons, D., and Chance, B., 1964. *Federation Proc.* 23:431.
Rakestraw, I., and Roberts, E., 1957, *Biochim. Biophys. Acta* 24:388.
Ramachandran, S., and Gottlieb, D., 1961. *Biochim. Biophys. Acta* 53:396.
Ramaiah, A., and Nicholas, D., 1964. *Biochim. Biophys. Acta* 86:459.
Raman, S., Sharma, B., Jayaraman, J., and Ramasarma, T., 1965. *Arch. Biochem. Biophys.* 110:75.
Razin, S., and Argaman, M., 1963. *J. Gen. Microbiol.* 30:155.
Razin, S., Argaman, M., and Avigan, J., 1963. *J. Gen. Microbiol.* 33:477.
Razin, S., Morowitz, H., and Terry, T., 1965. *Proc. Natl. Acad. Sci. U.S.* 54:219.
Razin, S., Tourtellotte, M., McElhaney, R., and Pollack, J., 1966. *J. Bacteriol.* 91:609.
Rebel, G., and Mandel, P., 1965. *Biochim. Biophys. Acta* 98:380.
Rebel, G., Sensenbrenner, M., and Mandel, P., 1964. *Bull. Soc. Chim. Biol.* 46:1113.
Redfearn, E., and Burgos, J., 1966. *Nature* 209:711.
Redfearn, E., and King, T., 1964. *Nature* 202:1313.
Redfearn, E., and Pumphrey, A., 1958. *Biochim. Biophys. Acta* 30:437.
Redfearn, E., Pumphrey, A., and Fynn, G., 1960. *Biochim. Biophys. Acta* 44:404.
Rees, M., and Nason, A., 1965. *Biochem. Biophys. Res. Commun.* 21:248.
Rees, M., and Nason, A., 1966. *Biochim. Biophys. Acta* 113:398.
Remsen, C., and Lundgren, D., 1963. *Bacteriol. Proc.* 33.
Repaske, R., 1954. *J. Bacteriol.* 68:555.
Repaske, R., 1958. *Biochim. Biophys. Acta* 30:225.
Repaske, R., 1961. In: *Abstract of Sectional Papers of Fifth International Biochemical Congress,* Vol. II (Russian edition), USSR Academy of Sciences Press, Moscow, p. 57. Available in English as: *International Congress of Biochemistry, 5th, Moscow, 1961. Biochemistry: Proceedings,* 9 Vols., Sissakian, N. M. (ed.), Pergamon Press, London.
Repaske, R., and Lizotte, C., 1965. *J. Biol. Chem.* 240:4774.
Rieske, J., and Zaugg, W., 1962. *Biochem. Biophys. Res. Commun.* 8:421.
Riklis, E., and Rittenberg, D., 1961. *J. Biol. Chem.* 236:2526.

Ris, H., and Plaut, W., 1962. *J. Cell Biol.* 13:383.
Ritchie, A., and Ellinghausen, H., 1965. *J. Bacteriol.* 89:223.
Robertson, J., 1959. *Biochem. Soc. Symp.* 16:3.
Robertson, J., 1961. In: Kety, S., and Elkes, J. (eds.), *International Neurochemical Symposium, 4th, Regional Neurochemistry: Proceedings,* Pergamon Press, London.
Robertson, J., 1964. In: Locke, M. (ed.), *Cellular Membranes in Development,* Academic Press, Inc., New York, p. 1.
Robinson, D., and Mills, R., 1961. *Biochim. Biophys. Acta* 48:77.
Robrish, S., and Marr, A., 1962. *J. Bacteriol.* 83:158.
Rodwell, A., and Abbott, A., 1961. *J. Gen. Microbiol.* 25:201.
Rose, I., and Ochoa, S., 1956. *J. Biol. Chem.* 220:307.
Rossi, C., Bielawski, J., and Lehninger, A., 1966. *J. Biol. Chem.* 241:1919.
Rossi, C., and Lehninger, A., 1964. *J. Biol. Chem.* 239:3971.
Rothblat, G., Ellis, D., and Kritchevsky, D., 1964. *Biochim. Biophys. Acta* 84:340.
Rothfield, L., and Horecker, B., 1964. *Proc. Natl. Acad. Sci. U.S.* 52:939.
Ruban, E., 1961. *Physiology and Biochemistry of Nitrifying Microorganisms,* USSR Academy of Sciences Press, Moscow.
Rudney, H., and Sugimura, T., 1961. In: Wolstenholme, G., and O'Connor, C. (eds.), *Quinones in Electron Transport,* Churchill, London, p. 211.
Russel, P., and Brodie, A., 1960. *Federation Proc.* 19:A–38.
Rýc, M., and Nermut, M., 1966. *Folia Microbiol. (Prague)* 11:95.
Ryter, A., 1965. *Ann. Inst. Pasteur* 108:40.
Ryter, A., and Jacob, F., 1963. *Compt. Rend. Acad. Sci.* 257:3060.
Ryter, A., and Landman, O., 1964. *J. Bacteriol.* 88:457.
Ryter, A., and Pillot, J., 1963. *Ann. Inst. Pasteur* 104:496.

Sadana, J., and McElroy, W., 1957. *Arch. Biochem. Biophys.* 67:16.
Sadana, J., and Morey, A., 1960. *Biochim. Biophys. Acta* 50:153.
Salem, J., 1964. In: Pullman, B. (ed.), *Electronic Aspects of Biochemistry,* p. 293.
Salem, L., 1962. *Can. J. Biochem. Physiol.* 40:1287.
Salton, M., 1957. *Second International Congress on Surface Activity,* Vol. 4, London, p. 245.
Salton, M., 1964. *The Bacterial Cell Wall,* American Elsevier Publishing Company, Inc., New York.
Salton, M., and Chapman, J., 1962. *J. Ultrastruct. Res.* 6:489.
Salton, M., and Netschey, A., 1965. *Biochim. Biophys. Acta* 107:539.
Sasaki, T., 1964. *J. Biochem.* 55:225.
Sato, R., 1956. In: McElroy, W., and Glass, B. (eds.), *Inorganic Nitrogen Metabolism,* Johns Hopkins Press, Baltimore, Maryland, p. 163.
Sazykin, Yu. O., 1965. *Biochemical Bases of the Action of Antibiotics on the Microbial Cell,* Nauka Press, Moscow.
Schachman, H., Pardee, A., and Stanier, R., 1952. *Arch. Biochem. Biophys.* 38:245.
Schaeffer, P., 1952. *Biochim. Biophys. Acta* 9:261.
Schatz, G., and Racker, E., 1966. *J. Biol. Chem.* 241:1429.
Scheide, O., McCondless, R., and Munn, R., 1964. *Nature* 203:158.
Schiels, D., Hovenkamp, H., and Colpa-Boonstra, T., 1960. *Biochim. Biophys. Acta* 43:129.
Schlegel, H., 1960. In: *Handbuch der Pflanzenforschung,* Vol. 5, Springer Verlag, Berlin, p. 649.
Schlessinger, D., 1963. *J. Mol. Biol.* 7:569.
Scholes, P., and King, H., 1963. *Biochem. J.* 87:10P.
Scholes, P., and King, H., 1965a. *Biochem. J.* 97:754.

Scholes, P., and King, H., 1965b. *Biochem. J.* 97:766.
Schulman, J., Pethica, B., Few, A., and Salton, M., 1955. In: Butler, J. A. V., and Katz, B. (eds.), *Progress in Biophysics and Biophysical Chemistry*, Vol. 5, p. 41, Pergamon Press, London.
Schwartz, A., and Perry, J., 1953. *Surface-Active Agents and Detergents* (Russian translation), Foreign Literature Press, Moscow. Originally published in English by (Interscience) Wiley.
Scocca, I., and Pinchot, G., 1963. *Biochim. Biophys. Acta* 71:193.
Scocca, I., and Pinchot, G., 1965. *Federation Proc.* 24(2) part 1:544.
Scott, P., 1965. *J. Biol. Chem.* 240:1374.
Sedar, A., and Burde, R., 1965a. *J. Cell Biol.* 24:285.
Sedar, A., and Burde, R., 1965b. *J. Cell Biol.* 27:53.
Segel, W., and Goldman, D., 1963. *Biochim. Biophys. Acta* 73:380.
Senez, J., 1962. *Bacteriol. Rev.* 26:95.
Seufert, W., 1965. *Nature* 207:174.
Shah, S., and King, H., 1965. *Biochem. J.* 94:13P.
Shethna, Y., Wilson, P., and Beinert, H., 1966. *Biochim. Biophys. Acta* 113:225.
Shibko, S., and Pinchot, G., 1961a. *Arch. Biochem. Biophys.* 93:140.
Shibko, S., and Pinchot, G., 1961b. *Arch. Biochem. Biophys.* 94:257.
Shinohara, C., Fukushi, K., and Suzuki, I., 1957. *J. Bacteriol.* 74:413.
Shockman, G., Kolb, J., Bakay, B., Conover, M., and Toennis, G., 1963.
Shol'ts, Kh. F., and Ostrovskii, D. N., 1965. *Lab. Delo* 6:375.
Sih, C., and Bennett, R., 1962. *Biochim. Biophys. Acta* 56:584.
Sinagoglu, O., and Abdulnur, O., 1965. *Federation Proc.* 24(2) part III:12.
Singer, T., 1963. In: Sumner, J., and Myrbäck, K. (eds.), *The Enzymes, Chemistry and Mechanism of Action*, Vol. 7, Academic Press, Inc., New York, p. 345.
Singer, T., and Lara, F., 1958. *Proceedings of the International Symposium on Enzyme Chemistry*, Maruzen, Tokyo, p. 203.
Singha, D., and Gaby, W., 1964. *J. Biol. Chem.* 239:3668.
Sisakyan, N. M., and Bekina, R. M., 1964. *Izv. Akad. Nauk SSSR, Ser. Biol.* 2:257; 3:396.
Sjöstrand, F., 1963a. *Nature* 199:1262.
Sjöstrand, F., 1963b. *J. Ultrastruct. Res.* 9:561.
Sjöstrand, F., and Elfin, L., 1964. *J. Ultrastruct. Res.* 10:263.
Skulachev, V. P., 1962. *The Ratio of Oxidation and Phosphorylation in the Respiratory Chain*, USSR Academy of Sciences Press, Moscow.
Skulachev, V., 1963. *Nature* 198:444.
Skulachev, V. P., 1964a. *Usp. Biol. Khim.* 6:180.
Skulachev, V., 1964b. *6th Intern. Congr. Biochem. Abstr. (New York)* X:758.
Slater, E., 1959. *Biochem. Soc. Symp.* 15:76.
Slater, E., 1961. In: Wolstenholme, G., and O'Connor, C. (eds.), *Quinones in Electron Transport*, Churchill, London, p. 161.
Slater, E., 1964. *Nature* 204:131.
Slater, E., and Cleland, K., 1952. *Nature* 170:118.
Slater, E., and Hülsman, W., 1959. In *Ciba Foundation Symposium on Regulation of Cell Metabolism*, Churchill Ltd., London, p. 58.
Smith, A., and Hansen, M., 1962. *Biochem. Biophys. Res. Commun.* 8:136.
Smith, A., and Hansen, M., 1964. *Biochem. Biophys. Res. Commun.* 15:431.
Smith, D., 1963. *J. Cell Biol.* 19:115.
Smith, E., and Margoliasch, E., 1964. *Federation Proc.* 23:1243.
Smith, L., 1954a. *Bacteriol. Rev.* 18:106.
Smith, L., 1954b. *Arch. Biochem. Biophys.* 50:255.
Smith, L., 1954c. *Arch. Biochem. Biophys.* 50:299.

Smith, L., 1955a. In: Colowick, S., and Kaplan, N. (eds.), *Methods in Enzymology,* Vol. 2, Academic Press, Inc., New York, p. 737.

Smith, L., 1955b. *J. Biol. Chem.* 215:847.

Smith, L., 1957. In: *Research in Photosynthesis.* (Interscience) Wiley, New York, p. 179.

Smith, L., 1959. *J. Biol. Chem.* 215:1571.

Smith, L., 1962. *Biochim. Biophys. Acta* 62:145.

Smith, L., 1963. In: Gunsalus, I., and Stanier, R., *Bacteria: A Treatise on Structure and Function,* Vol. II (Russian translation), Foreign Literature Press, Moscow, p. 360. Originally published in English by Academic Press, Inc., New York, 5 vols., 1960–1964.

Smith, L., and Minnaert, K., 1965. *Biochim. Biophys. Acta* 105:1.

Smith, L., and White, D., 1962. *J. Biol. Chem.* 237:1332.

Smith, P., 1963. *J. Gen. Microbiol.* 32:307.

Smith, P., 1964. *Bacteriol. Rev.* 28:97.

Smith, P., and Rothblat, G., 1960. *J. Bacteriol.* 80:842.

Snart, R., 1964. *Biochim. Biophys. Acta* 88:502.

Snoswell, A., 1966. *Biochemistry* 5:1660.

Sokawa, Y., 1965. *J. Biochem.* 57:706.

Sokawa, Y., and Egami, F., 1965. *J. Biochem.* 57:64.

Sokolova, G. A., and Karavaiko, G. M., 1964. *The Physiology and Geochemical Activity of Thiobacilli,* Nauka Press, Moscow.

Spiegelman, S., Aronson, A., and Fitz-James, P., 1958. *J. Bacteriol.* 75:102.

Srinivasan, V., and Halvorson, H., 1961. *Biochem. Biophys. Res. Commun.* 4:409.

Stanier, R., 1964. In: Gunsalus, J., and Stanier, R. (eds.), *Bacteria: A Treatise on Structure and Function,* 5 vols., Academic Press, Inc., New York.

Stanier, R., Gunsalus, I., and Gunsalus, C., 1953. *J. Bacteriol.* 66:543.

Stanier, R., and Van Niel, C., 1962. *Arch. Mikrobiol.* 42:17.

Stasny, J., and Crane, F., 1964. *J. Cell Biol.* 22:49.

Stephens, W., and Starr, M., 1963. *J. Bacteriol.* 86:1070.

Stoeckenius, W., 1959. *J. Biophys. Biochem. Cytol.* 5:491.

Stoeckenius, W., 1963. *J. Cell Biol.* 17:443.

Stoeckenius, W., 1964. *6th Intern. Congr. Biochem. Abstr. (New York)* VIII:603.

Stoeckenius, W., Schulman, I., and Prince, L., 1960. *Kolloid-Z.* 169:170.

Stolp, H., and Starr, M., 1965. *Ann. Rev. Microbiol.* 19:79.

Storck, R., and Wachsman, J., 1957. *J. Bacteriol.* 73:784.

Stouthamer, A., 1961. *Biochim. Biophys. Acta* 48:484.

Stouthamer, A., 1962. *Biochim. Biophys. Acta* 56:19.

Stove Poindexter, J., and Cohen-Bazire, G., 1964. *J. Cell Biol.* 23:587.

Straat, P., and Nason, A., 1965. *J. Biol. Chem.* 240:1412.

Strickland, E., and Benson, A., 1960. *Arch. Biochem. Biophys.* 88:344.

Suganuma, A., 1961. *J. Biophys. Biochem. Cytol.* 10:292.

Suganuma, A., 1965. *Ann. N. Y. Acad. Sci.* 128:26.

Sugimura, T., and Okabe, K., 1962. *J. Biochem.* 52:235.

Sutherland, J., 1963. *Biochim. Biophys. Acta* 73:162.

Sutherland, J., and Wilkinson, J., 1962. *Biochem. J.* 84:43P.

Sutton, D., and Starr, M., 1960. *J. Bacteriol.* 80:104.

Suzuki, I., 1965. *Biochim. Biophys. Acta* 104:359.

Taber, H., and Morrison, M., 1964. *Arch. Biochem. Biophys.* 105:367.

Tabor, C., 1962. *J. Bacteriol.* 83:1101.

Tabor, H., Tabor, C., and Rosenthal, S., 1961. *Ann. Rev. Biochem.* 30:579.

224   Literature Cited

Takagi, A., Kawata, T., Nanba, H., Inoue, T., Ueda, M., Abe, O., and Takeda, K., 1963. *J. Electronmicroscopy (Tokyo)* 12:277.
Takagi, A., Nakamura, K., and Ueda, M., 1965. *Japan J. Microscop.* 9:131.
Takagi, A., and Ueyama, K., 1963. *Japan J. Microbiol.* 7:43.
Takagi, A., Ueyama, K., and Ueda, M., 1963. *J. Gen. Appl. Microbiol.* 9:287.
Takahashi, K., Titani, K., and Minakami, S., 1959. *J. Biochem.* 46:1323.
Takeyoshi, N., 1961. *J. Biochem.* 49:240.
Tani, J., and Hendler, R., 1964. *Biochim. Biophys. Acta* 80:279, 307.
Taniguchi, S., Asano, M., Iida, K., Kono, M., Ohmachi, K., and Egami, F., 1958. *Proceedings of the International Symposium Enzyme Chemistry,* Maruzen, Tokyo, p. 238.
Taniguchi, S., and Itagaki, E., 1960. *Biochim. Biophys. Acta* 44:263.
Taniguchi, S., and Kamen, M., 1965. *Biochim. Biophys. Acta* 96:395.
Taniguchi, S., and Ohmachi, K., 1960. *J. Biochem.* 48:50.
Taniguchi, S., Sato, R., and Egami, F., 1956. In: McElroy, W., and Glass, B. (eds.), *Inorganic Nitrogen Metabolism,* Johns Hopkins Press, Baltimore, Maryland, p. 87.
Tchan, Y., Birch-Andersen, A., and Jensen, H., 1962. *Arch. Mikrobiol.* 43:50.
Temperli, A., and Wilson, P., 1958. *Experientia* 14:363.
Temperli, A., and Wilson, P., 1960. *Z. Physiol. Chem.* 320:195.
Temperli, A., and Wilson, P., 1962. *Nature* 193:171.
Theorell, H., 1947. *Advan. Enzymol.* 7:265.
Thompson, T., 1964a. Cited by Lehninger, A., 1964. *The Mitochondrion,* Benjamin Press, New York.
Thompson, T., 1964b. In: Locke, M. (ed.), *Cellular Membranes in Development,* Academic Press, Inc., New York, p. 83.
Thorne, K., 1964. *Biochim. Biophys. Acta* 84:350.
Thorne, K., and Kodicek, E., 1962. *Biochim. Biophys. Acta* 59:306.
Tikhonenko, A. S., and Bespalova, I. A., 1964. *Mikrobiologiya* 33:353.
Tikhonova, G. V., Skulachev, V. P., Lisenkova, L. L., and Doman, N. G., 1967. *Biokhimiya* (in press).
Tissières, A., 1951. *Biochem. J.* 50:279.
Tissières, A., 1954. *Nature* 174:183.
Tissières, A., 1956. *Biochem. J.* 64:582.
Tissières, A., 1961. In: Falk, J., Lemberg, R., and Morton, R. (eds.), *Hematin Enzymes,* Pergamon Press, Oxford, p. 218.
Tissières, A., and Burris, R., 1956. *Biochim. Biophys. Acta* 20:436.
Tissières, A., Hovenkamp, H., and Slater, E., 1957. *Biochim. Biophys. Acta* 25:336.
Tissières, A., and Slater, E., 1955. *Nature* 176:736.
Titani, A., Minakami, S., and Mitsui, H., 1960. *J. Biochem.* 47:290.
Tobari, J., 1964. *Biochem. Biophys. Res. Commun.* 15:50.
Tomasz, A., Jamieson, J., and Ottolenghi, E., 1964. *J. Cell Biol.* 22:453.
Tourtellotte, M., Jensen, R., Gander, G., and Morowitz, H., 1963. *J. Bacteriol.* 86:370.
Trudinger, P., 1958. *Biochim. Biophys. Acta* 20:211.
Trudinger, P., 1961. *Biochem. J.* 78:673.
Trudinger, P., 1964. *Biochem. J.* 90:640.
Tucker, R., 1960. *J. Gen. Microbiol.* 23:267.
Tyler, D., and Estabrook, R., 1966. *J. Biol. Chem.* 241:1672.
Tzagoloff, A., and MacLennan, D., 1965. *Biochim. Biophys. Acta* 99:476.

Unemoto, T., Hay-a-shi, M., and Miyaki, K., 1965. *Biochim. Biophys. Acta* 110:319.

Van Deenen, L., 1964. *6th Intern. Congr. Biochem. Abstr. (New York)* VII:553.
Van Deenen, L., 1965. *Phospholipids and Biomembranes,* Pergamon Press, London.

Van Deenen, L., and De Haas, G., 1963. *Biochim. Biophys. Acta* 70:538.
Van Deenen, L., and Van Golde, L., 1966. *Biochem. J.* 98:17p.
Van Demark, P., and Smith, P., 1964. *J. Bacteriol.* 88:122.
Vanderwinkel, E., and Murray, R., 1962. *J. Ultrastruct. Res.* 7:185.
Van Gool, A., and Laudelout, H., 1966. *Biochim. Biophys. Acta* 113:41.
Van Iterson, W., 1961. *J. Biophys. Biochem. Cytol.* 9:183.
Van Iterson, W., 1963. In: Gibbons, N. (ed.), *Recent Progress in Microbiology,* University of Toronto Press, Toronto, p. 2.
Van Iterson, W., 1965. *Bacteriol. Rev.* 29:299.
Van Iterson, W., and Leene, W., 1964a. *J. Cell Biol.* 20:361.
Van Iterson, W., and Leene, W., 1964b. *J. Cell Biol.* 20:377.
Van Tubergen, R., and Setlow, R., 1961. *Biophys. J.* 1:589.
Vennes, J., and Gerhardt, P., 1956. *Science* 124:535.
Verhoeven, W., and Takeda, Y., 1956. In: McElroy, W., and Glass, B. (eds.), *Inorganic Nitrogen Metabolism,* Johns Hopkins Press, Baltimore, Maryland, p. 159.
Vernon, L., 1956. *J. Biol. Chem.* 222:1035.
Vernon, L., 1958. *Proceedings of the International Symposium on Enzyme Chemistry,* 1957, Maruzen, Tokyo, p. 234.
Vernon, L., and Mangum, J., 1960. *Arch. Biochem. Biophys.* 90:103.
Vernon, L., Mangum, J., Beck, J., and Shafia, F., 1960. *Arch. Biochem. Biophys.* 88: 227.
Vernon, L., and White, F., 1957. *Biochim. Biophys. Acta* 25:321.
Vinnikov, Ya. A., 1964. In: *Molecular Biology,* Nauka Press, Moscow, p. 304.
Vishniac, W., and Trudinger, P., 1962. *Bacteriol. Rev.* 26:168.
Voelz, H., 1964. *J. Bacteriol.* 88:1196.
Voelz, H., 1965. *Arch. Mikrobiol.* 51:60.
Vorbeck, M., and Marinetti, G., 1964a. *Biochemistry* 4:296.
Vorbeck, M., and Marinetti, G., 1964b. *Federation Proc.* 22:1092.

Wachsman, J., and Storck, R., 1960. *J. Bacteriol.* 80:600.
Wadkins, C., and Lehninger, A., 1963a. *Federation Proc.* 22:1092.
Wadkins, C., and Lehninger, A., 1963b. In: Colowick, S., and Kaplan, N. (eds.), *Methods in Enzymology,* Vol. 6, p. 265.
Wadkins, C., and Mills, R., 1956. *Federation Proc.* 15:377.
Wagner, A., and Folkers, K., 1963. *Perspectives Biol. Med.* 6:347.
Wagner, A., and Folkers, K. (eds.), 1964. *Vitamins and Coenzymes.* (Interscience) Wiley, New York.
Wainio, W., 1960. *Trans. N. Y. Acad. Sci.* Series II 22:4.
Wainio, W., 1961. In: Falk, J., Lemberg, R., and Morton, R. (eds.), *Hematin Enzymes,* Vol. I, Pergamon Press, Oxford, p. 281.
Walter, H., and Eagon, R., 1964. *J. Bacteriol.* 88:25.
Waring, W., and Werkman, C., 1944. *Arch. Biochem. Biophys.* 4:75.
Warringa, M., and Guiditta, A., 1958. *J. Biol. Chem.* 230:111.
Warringa, M., Smith, O., Guiditta, A., and Singer, T., 1958. *J. Biol. Chem.* 230:97.
Watanabe, T., Murthy, P., and Brodie, A., 1965. *Federation Proc.* 24(2) part I:362.
Weber, M., Brodie, A., and Merselis, J., 1958. *Science* 128:896.
Weber, M., Hollocher, T., and Rosso, G., 1965. *J. Biol. Chem.* 240:1776.
Weber, M., and Rosso, G., 1963. *Proc. Natl. Acad. Sci. U.S.* 50:710.
Weber, M., Rosso, G., and Noll, H., 1963, *Biochim. Biophys. Acta* 71:355.
Weibull, C., 1953a. *J. Bacteriol.* 66:137.
Weibull, C., 1953b. *J. Bacteriol.* 66:688.
Weibull, C., 1953c. *J. Bacteriol.* 66:696.
Weibull, C., 1956. *Symp. Soc. Gen. Microbiol.* 6:111.

Weibull, C., 1957. *Acta Chem. Scand.* 11:881.
Weibull, C., Beckman, H., and Bergström, L., 1959. *J. Gen. Microbiol.* 20:519.
Weibull, C., and Bergström, L., 1958. *Biochim. Biophys. Acta* 30:340.
Weibull, C., Greenawalt, I., and Low, H., 1962. *J. Biol. Chem.* 237:847.
Weidel, W., and Pelzer, H., 1964. *Advan. Enzymol.* 26:193.
Weimberg, R., 1963. *Biochim. Biophys. Acta* 30:340.
Whatley, F., 1962. *Plant Physiol.* 37:8.
White, D., 1962. *J. Bacteriol.* 83:851.
White, D., 1963. *J. Bacteriol.* 85:84.
White, D., 1964. *J. Biol. Chem.* 239:2055.
White, D., 1965a. *J. Biol. Chem.* 240:1387.
White, D., 1965b. *J. Bacteriol.* 89:299.
White, D., 1966. *Antonie van Leeuwenhoek J. Microbiol. Serol.* 32:139.
White, D., Bryant, M., and Caldwell, D., 1962. *J. Bacteriol.* 84:822.
White, D., and Smith, L. *J. Biol. Chem.* 239:1932.
White, D., and Smith, L., 1964. *J. Biol. Chem.* 239:3956.
Whittaker, V., 1964. *6th Intern. Congr. Biochem. Abstr. (New York)* VIII:622.
Whittam, R., 1962. *Biochem. J.* 83:29P.
Whittam, R., Wheeler, K., and Blake, A., 1964. *Nature* 203:720.
Wildon, D., and Mercer, F., 1963. *Arch. Mikrobiol.* 47:31.
Williams, J., and Parsons, D., 1964. *6th Intern. Congr. Biochem. Abstr. (New York)* VIII: 670.
Wills, E., 1954. *Biochem. J.* 57:109.
Wimpenny, I., Ranlett, M., and Gray, C., 1963. *Biochim. Biophys. Acta* 73:170.
Wittenberger, C., and Repaske, R., 1961. *Biochim. Biophys. Acta* 47:542.
Wojtczak, L., Zaluska, H., and Drahota, Z., 1965. *Biochim. Biophys. Acta* 98:8.
Wolin, W., Wolin, E., and Jacobs, N., 1961. *J. Bacteriol.* 81:911.
Worcel, A., Goldman, D., and Cleland, W., 1965. *J. Biol. Chem.* 240:3399.
Worden, P., and Sistrom, W., 1964. *J. Cell Biol.* 23:135.
Wyss, O., Marilyn, P., Neumann, G., and Socolofsky, M., 1961. *J. Biochem. Biophys. Cytol.* 10:505.

Yakovlev, V. A., and Levchenko, L. A., 1964. *Dokl. Akad. Nauk SSSR* 159:173.
Yamada, Y., Iida, K., and Uemura, T., 1966. *Agr. Biol. Chem. (Tokyo)* 30:95.
Yamanaka, T., 1964. *Nature* 204:253.
Yamanaka, T., Kijimoto, S., and Okunuki, K., 1963. *J. Biochem.* 53:416.
Yamanaka, T., Kijimoto, S., Okunuki, K., and Kusai, K., 1962. *Nature* 194:759.
Yamanaka, T., and Okunuki, K., 1963a. *Biochem. Z.* 338:62.
Yamanaka, T., and Okunuki, K., 1963b. *Biochim. Biophys. Acta* 67:379.
Yamanaka, T., and Okunuki, K., 1964. *J. Biol. Chem.* 239:1813.
Yamanaka, T., Ota, A., and Okunuki, K., 1962. *J. Biochem.* 51:253.
Yamanaka, T.,Ota, A., and Okunuki, K., 1964. *6th Intern. Congr. Biochem. Abstr. (New York)* X:750.
Yamashita, S., and Ishikawa, S., 1965. *J. Biochem.* 57:232.
Yonetani, T., 1959. *J. Biochem.* 46:917.
Yoneya, T., and Adams, E., 1961. *J. Biol. Chem.* 236:3272.
Yudkin, M., 1962. *Biochem. J.* 82:40P.
Yudkin, M., 1966. *Biochem. J.* 98:923.
Yudkin, M., and Davis, B., 1965. *J. Mol. Biol.* 12:193.

Zaitseva, G. N., 1963. *Nitrogen and Phosphorus Metabolism of* Azotobacter *during Its Development,* Doctoral Dissertation.
Zaitseva, G. N., 1965. *Biochemistry of* Azotobacter, Nauka Press, Moscow.

Zaitseva, G. N., Agatova, A. I., and Belozerskii, A. N., 1961. *Biokhimiya* 26:338.
Zaitseva, G. N., Ngo Ke Syong, and Belozerskii, A. N., 1963. *Biokhimiya* 28:172.
Zarnea, G., 1963. *Microbiologie,* Bucharest.
Zavarzin, G. A., 1958. *Mikrobiologiya* 27:401.
Zavarzin, G. A., 1964. *Usp. Mikrobiol.* 1:30.
Zeng-Yi-shen, 1965. *Acta Biochim. Biophys. Sinica* 5:547.
Zeng-Yi-shen, and Chow Kwang-yü, 1965. *Acta Biochim. Biophys. Sinica* 5:509.
Zickler, F., 1965. *Z. Allgem. Mikrobiol.* 5:164.
Ziegler, D., and Doeg, K., 1959. *Arch. Biochem. Biophys.* 85:282.
Zvyagil'skaya, R. A., 1964. *An Investigation of Respiration and Oxidative Phosphorylation in Mitochondria from the Yeast Endomyces magnusii,* Candidate Dissertation.

Index

229